The Russian Natural
Gas 'Bubble'

Research by the Energy and Environmental Programme is supported by generous contributions of finance and professional advice from the following organizations:

Amerada Hess • Arthur D Little • Ashland Oil • British Coal
British Nuclear Fuels • British Petroleum • European Commission
Department of Trade and Industry • Eastern Electricity
Enterprise Oil • ENRON Europe • Esso • Exxon • LASMO
Mitsubishi • Mobil • National Grid • National Power
Nuclear Electric • Overseas Development Administration
PowerGen • Saudi Aramco • Shell • Statoil • St Clements Services
Texaco • Total • Tokyo Electric Power Company

The Russian Natural Gas 'Bubble'

Consequences for European Gas Markets

Jonathan P. Stern

THE ROYAL INSTITUTE OF
INTERNATIONAL AFFAIRS
Energy and Environmental Programme

First published in Great Britain in 1995 by
Royal Institute of International Affairs, 10 St James's Square, London SW1Y 4LE
(Charity Registration No. 208 223)

Distributed exclusively by
The Brookings Institution, 1775 Massachusetts Avenue NW,
Washington DC 20036-2188

A catalogue record for this book is available from the British Library.

Paperback: ISBN 0 905031 92 X

Printed and bound by Redwood Books Ltd.
Cover by Visible Edge.
Cover illustration by Andy Lovel.

Contents

4. The 'Gas Bubble' and Beyond: Scenarios to 2010

Tables

Figures

Maps

Boxes

Preface

Natural gas has come a long way since the 1960s and 70s, when it was considered as a rather limited premium fuel. Its use in Europe has risen steadily, and with it imports from Russia. In the early 1980s, western Europe was under pressure to restrict imports for the fear that these supplies might be used by the former USSR as a political weapon. In the 1990s, the new fear arose that the collapse of the USSR might interrupt supplies in other, less calculated ways. But the Russian natural gas industry itself has prospered whilst much around it has crumbled, and in this report Jonathan Stern presents the opposite thesis: that Europe will be faced with a glut of cheap Russian natural gas. The argument laid out with study is so simple and compelling that it may soon suffer the fate of being considered obvious. It is therefore worth remembering that at the time of this publication, many in the industry are still talking about the need for investments totalling tens of billions of dollars to open up new gas resources in Russia, and to construct associated pipelines. Stern's contention is that these developments are now largely irrelevant: the inevitable collapse of internal consumption means that plenty of gas can be derived from existing and satellite fields for many years. In going on to explore the implications of this, he points out that this is not just an internal issue for the Russians; it affects the terms on which gas can be made available for export in ways that has repercussions for European energy structures as a whole.

As one of the world's leading experts on the international gas business, on the European industry, and on Russian energy issues, Jonathan is uniquely well placed to carry out such a study. When he stepped down from being Head of RIIA's Energy and Environmental Programme, in June 1992, he commented that he wanted more time to research, as well as to enjoy life with two small children. This, Jonathan's third publication as an Associate Fellow, demonstrates the wisdom and value of that decision.

May 1995   Michael Grubb
                                   Head, Energy and Environmental Programme

Acknowledgements

Many people have contributed to the preparation of this study and have helped to add perspectives and correct errors and omissions as the text has developed. I would particularly like to thank those who attended the study group at Chatham House. The International Energy Agency was kind enough to invite me to participate in the mission to prepare its Energy Survey of Russia and this provided a valuable opportunity to gather material.

As far as individuals are concerned, thanks are due to Philip Hanson for providing me with statistical material which I would not otherwise have found. Anita Gardner at Gas Strategies was kind enough to change the maps a number of times as I discovered that fields and pipelines were in the wrong place. Javier Estrada, Arild Moe, Julian Bowden and Mike Parker read the final draft and made a number of helpful suggestions. Special thanks are due to Bertram Pockney, for help with statistics and many useful insights, and to Zuzana Princova, who helped me think through a great deal of the methodology in the early work for the study.

At Chatham House, Michael Grubb encouraged me to focus on a specific issue in the huge canvas of the Russian gas and electricity industries. Matthew Tickle brought the text through the necessary hoops as painlessly as possible. And Gillian Bromley did a splendidly thorough editing job.

Needless to say, I bear full responsibility for the final text.

May 1995   Jonathan P. Stern

Statistical Note

All the statistics in this study are expressed in Russian billion (thousand million) cubic metres or trillion (thousand billion) cubic metres, using the following abbreviations:

BCM = billion cubic metres;

TCM = trillion cubic metres.

For an approximate conversion from Russian BCM to European standard BCM, the Russian figures should be reduced by 6.9%.

For an approximate conversion from Russian BCM to tonnes of oil equivalent, the Russian figures should be reduced by 14.5%.

Prices are denominated in dollars per million British thermal units – $/mmbtu. $1/mmbtu is approximately equivalent to $5.8/barrel of crude oil, and also to $36.4/thousand cubic metres.

Summary and Conclusions

In the period up to 2010 Russian gas exports to Europe are likely to increase by 50–100% above 1994 levels. While many other studies have foreseen similar developments, the conclusions presented here are not based on the availability of future production from new high-cost fields on the Yamal Peninsula or the Barents Sea. The table overleaf shows that 1994 production capacity will be adequate for most scenarios of demand and export growth up to 2010 (scenarios A–C). Only if Russian demand returns to its 1990 level *and* exports to Europe double (scenario D) will significant increases in production be required. Production capacity can be maintained at 1994 levels by producing gas from deeper horizons of existing fields, and developing smaller satellite fields in the Nadym-Pur-Taz region. The only significant expansion of pipeline capacity which will need to take place will be from Central Russia and the Urals to export markets. Only as the industry approaches 2010 may new sources of high-cost gas, accessed through new high-cost transmission systems, be required; and even this is not certain, given Russia's options to supplement its own supplies with gas from Kazakhstan and Turkmenistan.

The evolution of the 'bubble'

These conclusions are based on a view of internal Russian demand for gas which sees a continuation of the contraction which has already taken place since 1990. It is suggested here that Russian gas demand will fall to around 320 BCM per annum by 2000 and in 2010 is likely to lie in the range 350–400 BCM (roughly equivalent to demand levels in 1994 and 1990 respectively).

A 'bubble' of gas production capacity – amounting to more than 30 BCM in 1994 – remained shut in (i.e. not produced) because of lack of markets,

Russian gas balance: scenarios to 2010 (BCM)

	1994 (est.)	2000	2010 A	B	C	D
Exports[a]						
to FSU	79	80	80	80	80	80
to Europe	106	140	150	200	150	200
Imports	4	—	—	50	50	50
Demand	359[b]	320	350	350	400	400
Pipeline fuel[c]	58	57	60	60	60	67
Production						
actual	607	597	640	640	640	697
capacity	640	640	640	640	640	697
'Bubble'[d]	33	43	—	—	—	—

[a] FSU means former Soviet republics; Europe means Central/Eastern and OECD Europe.
[b] This figure does not include net additions to storage of around 9 BCM.
[c] Approximately 9.5% of production.
[d] Capacity which remains shut in for lack of markets.

domestic and foreign. Even with exports to Europe increasing by 40% over the next five years, this bubble is likely to increase to around 40 BCM by the end of the year 2000. Approaching 2005 the bubble is likely to deflate, but the timing of the deflation depends critically on trends in internal Russian gas demand – specifically government attitudes towards non-payment by Russian consumers – which in turn depend on the pace and extent of economic reform within the country.

The expansion of transmission capacity

Gazprom has serious intentions to expand transmission capacity to European countries from 115 BCM in 1994 to 140 BCM in 2000, and further to 200 BCM in 2010. Actual deliveries to European countries will probably expand more slowly than this because of difficulty in finding markets: for this reason, Russian gas exports to Europe in 2010 could be restricted to 150 BCM. In the short term, exports will be driven not by the availability of gas but by the availability of transmission capacity and markets. By 2000, new transmission capacity – made available principally by the creation of the new Belarus–Poland ('Yamal') corridor – will provide the means by which large additional volumes will reach Europe. Although the cost of creating this new capacity will be significant, exports will be profitable at 1994 European border prices.

The consequences for European gas markets

The expansion of Russian exports foreseen here is significant in itself, but it is the 'bubble' of 30–40 BCM – roughly equivalent to 10% of current (East and West) European gas demand – which is likely to:

- exert downward pressure on gas prices in European markets and accelerate the development of gas-to-gas competition;
- accelerate the development of liberalized access to transmission networks;
- retard substantially the development of higher-cost gas supplies within and around Europe, particularly deep-water Norwegian gas, greenfield liquefied natural gas (LNG) projects and Middle East pipeline gas;
- refocus the attention of European governments and companies on the security of Russian gas supplies, particularly in the light of problems in relations with Ukraine.

Any simplistic conclusion that large volumes of Russian gas will be 'dumped' on European markets at very low prices is almost certainly wrong. Gazprom (and its export division Gazexport) is too experienced in the European gas business to behave so foolishly, unless directed to do so by higher political authorities. Nevertheless, it is highly unlikely that Gazprom's new joint-venture marketing companies will succeed in creating sufficient export markets for Russian gas, sufficiently quickly to exhaust the potential of the bubble, if they insist on sticking to 'traditional' contractual terms: long term, take-or-pay contracts with gas priced against alternative fuels. If the bubble persists, as foreseen in this study, the temptation to move shorter-term, flexible packages of gas into the market at extremely competitive prices will become very great. Such developments are likely to be viewed as an increasingly serious threat by all other parties attempting to develop alternative supplies for European markets; and with increasing interest by consumers – particularly large consumers – throughout Europe.

Economic and political uncertainties

Western studies of Soviet and Russian gas (and oil) export potential have periodically predicted the likelihood of massive surpluses or massive shortfalls of supply, leading to glut or crisis on world markets. Thus far, none of these studies has been proved correct, for a mixture of economic, commercial and political reasons. This study may end up sharing the fate of those that have gone before, given the huge economic and political uncertainties of trying to forecast the present Russian situation 15 years ahead.

From an *economic* standpoint, the projections put forward here clearly depend on a relatively small difference between two relatively large numbers: gas supply and gas demand in Russia. A small percentage change in either number could rapidly eliminate the surplus foreseen in these projections. On the supply side, a faster than expected decline in production from existing fields or a slower than expected opening up of new production capacity could give rise to a fall in availability; on the demand side, the Russian economy could stage an outstanding recovery. Yet despite the many uncertainties in the Russian economic situation over the next decade, all the demand-side indicators – macroeconomic decline, enforcement of payment or disconnection of non-paying customers, contraction of energy-intensive industry, conservation and efficiency measures, replacement of inefficient plant, moves towards market pricing – point in the same direction.

In 1994 payment enforcement was the outstanding issue. In that year, 'real gas demand' – i.e. volumes of gas for which consumers paid officially decreed prices – in Russia and for Russian gas in former Soviet republics was around 205 BCM, while 'gas deliveries' to those customers approached 440 BCM. Thus an alternative interpretation of the 1994 gas bubble would be the 235 BCM which was delivered but not paid for, plus the 33 BCM of additional production capacity which could have been delivered had markets existed. However, the really interesting calculation would be the 'real' gas bubble, i.e. the volume of gas which would have been available if suppliers had demanded prompt payment at agreed (official) prices, and disconnected customers were unable or unwilling to pay. Such a figure can only be guessed: perhaps 50 BCM, perhaps larger. But the size and likely duration of the real gas bubble is the principal reason why this author finds it extremely difficult – short of a return to central planning in Russia – to construct a scenario in which Russian gas demand regains its 1990 level before 2010.

Yet this reasoning is strongly dependent on *political* factors, in particular the economic policy of future Russian governments. If the latter are content either to allow the current level of non-payment to continue, or to return to the pricing system of the Soviet period which did not recognize real costs, then gas demand could again increase strongly. However, it is difficult to see how these policies can be followed at the same time as market-oriented reforms. As long as Russian governments remain committed to market-oriented economic reform – however weakly and on however long a time scale – demand for energy, including gas, is unlikely to rise significantly over the next 15 years.

Thus, if the Russian gas bubble fails to reach Europe, the reasons are more likely to be political than economic or commercial. While a catastrophic political/military event within Russia or Ukraine – or between the two countries – is possible, it remains unlikely. More likely, given the unstable political situation in Russia, is the prospect of a future government adopting authoritarian domestic policies, illiberal economic and foreign trade policies, and neo-imperialistic foreign policies (both towards former Soviet republics and conceivably towards Central/East European countries).

Security of supply

Such events would increase concerns about security of Russian gas supply which already exist because of problems with transit through third countries. While these problems currently centre on Ukraine, it should not be assumed that Russian relations with the Czech and Slovak Republics, Belarus or Poland will be trouble-free over the next 15 years. Security problems are likely to serve as a convenient argument for European governments to use in order to protect their domestic gas (and other energy) industries. The political aspects of Russian gas exports will again come to the fore, as they have periodically over the past 25 years. The major issue in the post-Soviet era will be whether the security problems associated with Russian gas will become sufficiently serious in reality – or the prospects of such problems sufficiently well argued by their proponents – to prevent very large quantities of attractively priced gas moving into Europe within a relatively short period. Unless halted by adverse political/military events, the pressures caused by a large and rapid increase in Russian gas exports will serve as an important, perhaps decisive, driving force towards liberalization and competition in European gas (and other energy) markets over the next decade.

Introduction

The principal motivation for this study arose from a conviction that the traditional way in which the outlook for Russian gas exports has been viewed in Europe is fundamentally misconceived and is itself a hangover from the Soviet era. Up to the present, virtually all forecasts of Russian gas trade have been supply- and resource-driven. Recently data have become available which allow a more detailed examination of the demand side of the gas balance to be made. This is perhaps the area where the present study can make a new contribution to published literature.

The study is organized in four chapters.

Chapter 1 looks at resources, current and future supply issues and the costs of various production options.

Chapter 2 looks at Russian gas demand and attempts to make some projections based on a methodology which focuses on likely developments in individual end-use sectors and regions of the country.

Chapter 3 discusses export issues and the transition from Soviet to post-Soviet arrangements, dividing markets between former Soviet republics and European countries. This chapter also looks at availability of pipeline transportation capacity to Europe.

Chapter 4 looks at scenarios for gas exports in two time-frames: up to 2000 and up to 2010. It points to some major potential impacts on European gas markets of what is seen as the emerging Russian gas bubble, and the likely responses of European companies and governments, particularly in terms of security of supply concerns. The chapter concludes with some observations on Gazprom's likely strategy in the domestic and foreign markets which it serves.

Quantitative scenarios are put forward in order to allow readers to place a range of values on individual components of the emerging Russian gas balance. By this means, the study is intended to assist the reader's own projections in the current situation of economic and political uncertainty. Nevertheless, if the analysis presented here is directionally correct, the impact of Russian gas on European gas and energy markets, where it is already a major factor – and promises to play an even larger role in the future – could be profound.

Chapter 1

The Supply Side: Reserves, Field Development and Costs

1.1 The resource base

At the beginning of 1994, Russia had a proven gas reserve base of 48–9 TCM, which amounted to around 35% of world proven reserves.[1] While Russian reserves undoubtedly dominate the resource base of the former Soviet Union, very considerable gas potential also exists in the Central Asian states of Turkmenistan (with proven and probable reserves of 2.8 TCM), Uzbekistan (1.9 TCM) and Kazakhstan (1.8 TCM).[2] The current and potential inter-relationships between Russian and Central Asian gas reserves will become apparent later in the study (Chapters 3 and 4).

Around three-quarters of proven Russian gas reserves are located in north-western Siberia in some 20 'unique' fields located in the Nadym-Pur-Taz region and on the Yamal Peninsula (see Map 1.1).[3] More than 80% of current Russian production comes from three production associations in Siberia, based around three multi-TCM fields: Medvezhe, Urengoy and Yamburg.

The potential resource base is colossal. An extensive, but by no means complete, listing of Russian gas fields as of 1992 arrived at the breakdown in Table 1.1. According to Gazprom data, 68 fields are in production

[1] The Chairman of Gazprom gave the figure of 49 TCM in his presentation to the 1994 World Gas Conference in Milan. R. I. Vyakhirev, 'The Russian Joint Stock Company Gazprom: Problems and Perspectives'.
[2] Jean Christophe Fueg, 'The Gas Industry of the Southern FSU', in *Gas in the Former Soviet Union*, *Petroleum Economist*, Special Report, September 1994, pp. 24–6. Much larger reserve estimates are available for these countries. This source gives as additional 'estimated undiscovered reserves' for Turkmenistan, 3 TCM; for Uzbekistan, 2.4 TCM; and for Kazakhstan, 6.2 TCM. Another article in the same publication gives a figure of 21 TCM for the total Turkmen resource base.
[3] R.I. Vyakhirev, 'The Russian Gas Industry in the Context of the Russian and World Economy', paper presented to the conference on 'Natural Gas: Trade and Investment Opportunities in Russia and the CIS', Royal Institute of International Affairs, London, 13–14 October, 1994 (henceforth RIIA Conference).

Table 1.1 Categories of Russian gas fields

Field size[a]	In operation	Developed	Under appraisal	Undeveloped
Small	79	8	25	5
Medium	21	17	28	10
Large	23	14	62	48
Unique	9	6	7	6
Total	132	45	122	69

[a] All fields with gas reserves including: gas only, oil and gas, gas and condensate. While the size of fields is not defined in the source, a reasonable guess might be: small = up to 50 BCM, medium = 50–100 BCM, large = 100–500 BCM, unique = more than 500 BCM.
Source: calculated from VNIICTEP data in: *Gas in the Former Soviet Union*, *Petroleum Economist* Special Report, September 1994, pp. 68–86.

accounting for 17.9 TCM of reserves. While the grand total of fields in Table 1.1 is 368, at least half of these would not be included under the heading of proven reserves, but rather in the larger Gazprom definition of potential reserves amounting to 212 TCM.[4]

1.2 The organization and corporate culture of the industry

Gazprom controls almost the entire Russian gas industry as far as the 'city gate'. In 1993 the company produced 93.4% of Russian gas (i.e. all gas produced by the gas industry), and it owns all the high pressure transmission lines.[5] Gazprom is the direct descendant of the Soviet Ministry of the Gas Industry, and retains many of the same people in senior positions.[6] Still run

---

[4] Ibid. For a very useful guide to the resource base see John D. Grace, 'Russian Gas Resource Base: Large, Overstated, Costly to Maintain', *Oil and Gas Journal*, 6 February, 1995, pp. 71–4.
[5] It does not, however, control gas distribution, which is carried out by distribution companies under the umbrella of RosGasifikatsiya. The remaining 6.6% of gas was produced by oil production companies (of which the largest share was associated gas from the West Siberian oilfields under the control of Rosneft Associations). Furthermore, despite its almost total control of current production, Gazprom owns only 50% of the 49 TCM of proven reserves. 'Gazprom Study Cracks Open Secrets of the Gas Giants', *World Gas Intelligence*, 27 January, 1995, p. 5.
[6] Useful background can be found in Arild Moe, *The Organisation of the Russian Gas Industry* (Oslo, Norway: Fridtjof Nansen Institute, 1994).

Gazprom has undoubtedly benefited from close links with the present Russian Prime Minister Viktor Chernomyrdin, who was Soviet Minister of the Gas Industry and was responsible for the creation of Gazprom first as a 'Kontsern' and then as a joint stock company. While Chernomyrdin remains in a politically powerful position, no serious changes in either the leadership or the monopoly position of Gazprom are likely. Were Chernomyrdin to withdraw or be ousted from the political scene, this situation could change.

The corporate culture of Gazprom, evolved in the Soviet era, remains heavily oriented towards the upstream development of a small number of 'megaprojects': multi-TCM fields with multiple strings of long-distance, large-diameter pipelines. The issue of 'demand projections', as that term would be understood in the West, barely featured in the minds of the former Soviet gas planners. Gas 'sales', involving exchange of gas for money, were much less important than detailed allocation of production to different markets in the former Soviet Union. Only in the context of exports to Europe – particularly OECD Europe – were revenues the paramount consideration. As we shall see in Chapter 2, this outlook has begun to change. The Soviet-type, centrally planned, supply-driven strategy has in any case come under severe pressure in an environment where Gazprom is required to provide all capital investment from its own resources, or through foreign investment.

1.3 Gas production and field development

During the past 25 years of Soviet and Russian gas development hundreds of billions of cubic metres of gas per year have been brought to the west of the country from four multi-TCM fields by means of multiple strings of 56 inch pipeline, several thousand kilometres in length (see Map 1.2). The Medvezhe field was brought into production in 1972, followed by Urengoy in 1978 and Yamburg in 1986. The two latter fields were sufficiently large to warrant the building of six 56 inch pipelines from each field to the west of the country, one of which was dedicated to exports. The Orenburg field (the only major non-Siberian field to have been brought into production in the modern era) came on-stream in 1979, with a pipeline exporting gas to Central/East European countries. In the early 1980s it was decided that, after Yamburg, the next major increment in Russian production would come from the Yamal

Map 1.2 Major Russian pipeline systems

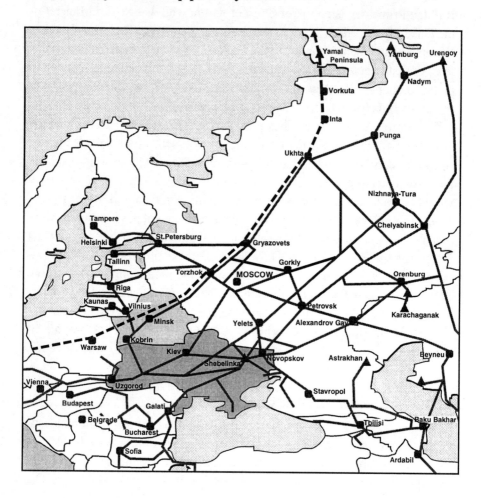

Source: Gas Strategies, London, 1995.

Table 1.2 Natural gas production in Russia, 1988–1994 (BCM)[a]

	1988	1991	1992	1993	1994[b]
Russia total	589.7	642.9	640.4	617.6	607.3
Gazprom production	546.7	601.6	602.7	577.7	570
West Siberia	475.4	533.3	549.8	535.5	
Nadymgazprom	73.3	68.9	69.1	68	64.3
Yamburggazodobycha	84.5	166.8	178.2	173	179.3
Urengoygazprom	298.8	282.8	287.6	281	
Surgutgazprom	18.8	14.8	14.9	14	
East Siberia	5.1	4.9	5.0	5.2	
Outside Siberia	66.2	63.4	47.9	37.0	
Orenburggazprom	45.9	48.0	36.4	27.1	
Astrakhangazprom	4.8	3.3	2.9	2.6	
Severgazprom	8.3	5.2	4.9	4.4	
Non-Gazprom production					
gas from oil production	43.1	41.3	37.7	39.9	
including associated gas	37.9	35.1	32.0		
non-associated gas	5.2	6.2	5.7		

[a] Totals may not add due to rounding.
[b] Preliminary figures.
Sources: David Wilson, *CIS and East European Energy Databook*, Eastern Bloc Research, 1994, table 33; *Gas in the Former Soviet Union*, *Petroleum Economist* Special Report, September 1994, p.8.

Peninsula, where 25 fields containing 10.2 TCM had been discovered. A more detailed look at existing and potential field development suggests some alternative options for the future.

Existing fields and satellites
Table 1.2 shows the trends in production from the major gas production associations since 1988. The Medvezhe field (represented by Nadymgazprom in the table) has been in decline for some years, and although the association's production seems relatively stable, Medvezhe has already produced a relatively high percentage of its initially recoverable reserves and new fields are compensating for the decline. Production at Urengoy peaked around 300 BCM in the mid-1980s and although it has fallen by about 10% since that time, it appears to have been used as the

'swing' field.[10] However, as at Medvezhe, satellite fields are beginning to compensate for Urengoy's decline. Production at Yamburg has increased steadily in the 1990s, but has remained below its design plateau level of 207 BCM per year because of lack of investment.

The major issue for the production associations at Urengoy and Nadym over the next 10–15 years is whether satellite fields can be developed sufficiently quickly to compensate for decline in the major fields. It is a matter of some guesswork whether such decline – particularly at Urengoy – will be gradual or rapid. (Some aspects of resource depletion at Urengoy and Yamburg are discussed in more detail in Box 1.1.) An argument in favour of gradual decline would be that since production at both Urengoy and (particularly) Yamburg has fallen during the period 1992–4 (rather than being increased, as had been anticipated), the pace of resource depletion – and tendency towards over-production – has been lessened. Against those arguments must be set the lack of investment and shortage of materials which may prevent production capacity being developed and/ or maintained.[11] It is very difficult to try to generalize from experience elsewhere because these are truly 'unique' fields. Perhaps the only long-term experience with a field of comparable size is at Groningen in the Netherlands, where production has been maintained at a higher level for a longer period than had been anticipated at the start of development. However, both the physical and the commercial environment in which that field is being operated are so completely different from those of Siberia that few comparisons are useful.

A reasonable generalization might be that, at current rates of production, the shallow, low-cost (Cenomanian) gas will have been largely exhausted at Urengoy around 2005, and Yamburg around 2010. Anticipation of this type of decline in production might indicate a pressing need to hurry on to the Yamal Peninsula fields. However, this tends to overlook gas in the deeper horizons of Urengoy and Yamburg and, possibly more important, smaller fields located relatively close to existing infrastructure (see Map 1.1).

---

[10] By 'swing field' is meant the field where production is either raised or lowered depending on short term demand conditions.

[11] *Eastern Bloc Energy*, August 1994, p. 4, quotes the Head of Construction at Yamburg as saying that reserves are being lost due to failure to maintain reservoir pressures because the head compressor stations have not been built on time.

Box 1.1 The timing of production decline at the Urengoy and Yamburg fields

Detailed technical aspects of production at Urengoy and Yamburg are beyond the scope of this study. However, the timing and speed of future production decline at these two fields will be the critical factor in planning the future development of Russian gas production.

The key issue is the extent to which the shallow – relatively easy to extract – Cenomanian reserves have been depleted, requiring the development of deeper, more technically challenging, and therefore more costly, Neocomian and Valanganian horizons. Dienes is extremely pessimistic about production prospects, particularly at Urengoy, asserting that: 'Ten percent of estimated reserves for the whole field and 30% in some parts of the field may be non-existent. No further increases in extraction are possible...the deposit will begin to decline as early as 1996 with sharp decreases in well productivity.'[a]

He notes that around 90% of gas produced from both Urengoy and Yamburg is from Cenomanian horizons and quotes an interview with Zhabrev in which the latter suggests that well yields from production from Valanganian horizons are much lower than from Cenomanian wells and cannot be expected to exceed 40 BCM per year.[b] Grace on the other hand suggests a much larger production potential by referring to an additional 10 TCM of reserves in place in the deeper horizons.[c]

Grace asserts that production of Cenomanian gas at Urengoy has already begun to decline and also emphasizes the stratigraphic complexity and greater expense involved in producing gas from the deeper horizons.[d] Grace's work could lead to the conclusion that instead of moving to the deeper horizons of Urengoy and Yamburg, the industry would be better advised to concentrate on development of satellites.

A rough computation reveals that, since the start of production, some 3400 BCM of gas have been produced at Urengoy and 750 BCM from Yamburg. Cenomanian reserves were estimated in 1980 at 6.2 TCM for Urengoy and 3.4 TCM for Yamburg.[e] Grace's rule of thumb that no more than 80% of quoted reserve figures should be considered recoverable would suggest that nearly 70% of recoverable Cenomanian reserves have been produced at Urengoy, but less than 25% at Yamburg. If this reasoning is correct, the suggestion of imminent rapid decline may have some force at Urengoy where, at the rates of extraction of the early 1990s, only around seven years of Cenomanian reserves may remain, although this might be somewhat extended by satellites (such as North Urengoy). The downward drift in Urengoy production will therefore certainly continue, although both Dienes and Grace suggest that it would be possible to increase recovery from Cenomanian reservoirs by using additional compression. At Yamburg, by contrast, at least 10 years of Cenomanian reserves exist, even at higher rates of production than those of the early 1990s. Only if the rumours that overly rapid production rates early in the development of the fields have significantly reduced ultimate recovery are to be believed would these time-scales be seriously shortened.

[a] Leslie Dienes, Istvan Dobozi and Marian Radetski, *Energy and Economic Reform in the Former Soviet Union* (New York: St Martins Press, 1994), p. 56; [b] Ibid., p. 223, n.16; [c] John D. Grace, 'Cost Russia's Biggest Challenge in Maintaining Gas Supplies', *Oil and Gas Journal*, 13 February, 1995, pp. 79–81; [d] Ibid.; [e] I. P. Zhabrev, *Gazovye i Gazokondensatnye Mestorazhdenie* (Moscow: Nedr, 1983), pp. 130–33, table 32.

Table 1.3 gives an indication of the production potential for some of these satellite fields. Grace suggests that in addition to Zapolyarnoye (which is arguably too large to be considered a satellite), there is a group of seven fields within 150 km of Urengoy with a total production potential of 140 BCM per year.[12] This figure is somewhat below the total in Table 1.3 which, including Zapolyarnoye, would give a total production of 272 BCM per year. These are however only a small fraction of the potential satellite fields. The standard Soviet source on gas reserves lists 52 gas fields in north-western Siberia, including the Taz, Gydan and Yamal Peninsulas.[13] More recent maps show 67 fields in the Nadym-Pur-Taz region located relatively close to existing infrastructure. Four are within 100 km of Medvezhe, 23 are within 150 km of Urengoy and 10 are within 150 km of Yamburg.[14]

Gazprom has suggested (see Table 1.4 below) that the annual production potential of the Nadym-Pur-Taz region is 740–755 BCM, which is more than 200 BCM higher than the current west Siberian production of 535 BCM and around 140–150 BCM higher than total Russian production in 1994.[15] This gives an indication of the potential for offsetting decline at the older fields, given also that in addition to those satellite fields close to existing production (listed in Table 1.3), there are plenty of smaller fields which could be opened up.[16] The task over the next 15 years may be to replace the current production from shallow horizons of existing fields either with

[12] John D. Grace, 'Cost Russia's Biggest Challenge in Maintaining Gas Supplies', *Oil and Gas Journal*, 13 February, 1995, pp. 79–81. Grace includes South Russkoye in his group of seven satellites, but does not include either Yety Purovsk or Kharvutinskoye.
[13] I. P. Zhabrev, *Gazovye i Gazokondensatnye Mestorazhdenie* (Moscow: Nedr, 1983), Figure 41, p. 128.
[14] Near Medvezhe: Pangodinsk, Kushelev, Nyda and Sandibiu; near Urengoy: Pestovoy, Yubilyeynoye, Yamsovey, Yevoyakha, E. Urengoy, N. Yesetin, S. Urengoy, Samburg, S. Samburg, N. Urengoy, S. Pyrey, Salekhart, Taz, Gaz-Sale, Zapolyarnoye, Russko Bechen, Russkoye, W. Zapolyarnoye, Neponyatnoye, S. Russkoye, N. Yesetin, Yaroyakh, Beregovoye; near Yamburg: Parusovoye, Kharvutinskoye, Aderpayuta, Semakov, Tota-Yakha, Antipalyutinsk, Khalmerpayuta, Minkhov, Nakhodkinsk, Yurkharov.
[15] B. V. Budzulyak. 'Upgrading and Retrofitting: Future Developments', in *Gas in the Former Soviet Union 1993*.
[16] See note 14 above.

Table 1.3 Siberian satellite field development

Field	Likely production level (BCM)[a]
Komsomolskoye[b]	30
Yubilyeynoye[b]	15
Yety Purovsk	10
Yamsovey	25
Kharvutinskoye	30
West Tarkosalinskoye	15
East Tarkosalinskoye	15
Gubkinskoye	32
Zapolyarnoye	100

[a] Forecast annual plateau production for 5–7 years.
[b] In production.
Source: Gazprom data, especially the paper by A. Pushkin presented to conference on 'Natural Gas: Trade and Investment Opportunities in Russia and the CIS', Royal Institute of International Affairs, London, October 1994.

production from shallow horizons of smaller fields, or with production from deeper horizons of existing fields. A mixture of both is the most likely outcome. From a resource perspective, therefore, there is no doubt that the region will be able to maintain its current production capacity in the period up to 2010. The major issue will be the length of time and cost needed to bring this new capacity on-stream.

The Yamal fields
As noted above, throughout the final decade of the Soviet era it was indicated that the next major Siberian gas development would be the 'Yamal' project. The majority of proven Yamal reserves are contained in three big fields: the Bovanenko/Kharasavey/Kruzenshtern cluster located in the middle of the Peninsula on the western shore (see Map 1.1). The original plan was to replicate the Urengoy and Yamburg model by building six 56 inch (1420 mm) pipelines running from the Bovanenko/Kharasavey field complex to link with the existing northern pipeline corridor at Ukhta. During the 1980s, much debate took place with local Yamal-Nenets ethnic peoples concerned about environmental and habitat disturbance which might be caused by the

pipeline development on the peninsula.[17] As a result, it was believed that the decision had been taken for the pipeline corridor from the fields to Ukhta to cross the Gulf of Baidarat instead of proceeding south down the Peninsula. Reports in late 1994 suggested that this was still the favoured route, but a final decision has yet to be taken.

From Ukhta, the domestic lines are intended to service the north west and centre of the country, while the export lines would follow the existing Northern Lights system through Belarus, branching at Brest into Poland and thence into Germany. In 1994 Gazprom was still suggesting that production from the Yamal fields would reach 170 BCM per year (i.e. requiring six pipelines), but the original plan to start production from Bovanenko by 1997 has been pushed back, and over the past two years the domestic phase has been scaled down considerably. By late 1994 Gazprom had indicated that the six projected pipelines had been reduced to three, running between the Peninsula and Torzhok, from where two export lines will continue to Belarus and Poland.[18] Current attention is being devoted to the export lines which, when fully operational, are intended to carry 67 BCM of gas per year (we shall return to this issue in Chapter 3).

The Barents and Kara Sea fields

The Shtokmanovskoye field in the Barents Sea, discovered in the late 1980s with reserves estimated at up to 3 TCM, provides an alternative production option to the Yamal Peninsula. But it is generally accepted – by, among others, Gazprom – that foreign technology and assistance will be needed to develop Shtokmanovskoye, which is capable of producing 50–100 BCM

---

[17] The original debate was defined in terms of the relative environmental disruption which would occur in the Gulf of Baidarat region, as against 'tearing up' a corridor the length of the Peninsula. In addition, the difficult ice conditions in the Gulf of Baidarat appeared to pose a technical problem. At one point it seemed that the Yamal-Nenets peoples might have the power to veto the entire project. It may be altogether too cynical to note that the issue seemed to have been resolved at the same time as the decision to award the Yamal-Nenets peoples a small but specific share-holding in the Gazprom privatization.

[18] V. I. Rezunenko, 'A Yamal Project for the Future', in *Gas in the Former Soviet Union 1994*, p. vi. This article suggests three 56 inch pipelines between the border and the Gulf of Baidarat. The crossing of the Gulf would be carried out using four 48 inch lines. More technical details are contained in *Proekt Yamal-Evropa* (Moscow: Gazprom, 1994).

per year for 25 years.[19] Institutional problems in the early 1990s set back this development,[20] but the Rosshelf-Gazprom consortium which was awarded the rights to develop the fields appears to have resumed cooperation with the original foreign joint venture partners. The Rusanovskoye and Leningradskoye fields in the Kara Sea, in the offshore regions of the Yamal Peninsula, with reserves of 5 TCM provide yet more potential to augment the already huge onshore finds. However, given the resource wealth which has already been identified onshore, production from these offshore discoveries may be some decades away.

1.4 Costs of production and transmission

The traditional Soviet approach to resource development rarely involved any public discussion of the costs involved, let alone the relationship of expected revenues from the gas sales to these costs. In the West, much speculation has always surrounded the commercial aspects of former Soviet gas projects. Western studies have periodically suggested that, calculated on the basis of market economics, none of the giant Siberian projects would have been profitable. However, it is very difficult to carry out this type of analysis since many of the detailed operational costs are not known, or are denominated in roubles at rapidly depreciating exchange rates. Little information on this issue is in the public domain, although it is known that internal Gazprom accounting allows for a system of transfer pricing along the gas chain, based on estimates of costs. As far as the latter are concerned, the main issue is that of maintenance and refurbishment of, particularly, the pipeline network.

More detailed analysis of Russian and Western estimates of gas production and transmission costs can be found in the Appendix to this chapter. The general conclusion from incomplete, inconsistent and questionable data is that the current cost of gas at the city gate, or export border, is around $1 per

---

[19] 'Rosshelf and the Barents Sea', *Eastern Bloc Energy*, August 1994, pp. 4–5; Grace, 'Cost Russia's Biggest Challenge', puts peak production at Shtokmanovskoye somewhat higher at 125 BCM per year.

[20] The basic institutional/political problem was that Gazprom was not included in the initial joint venture. For background see Arild Moe, 'The Energy Sector of the Barents Region', *International Challenges*, vol. 12, no. 4, 1992, pp. 57–68.

mmbtu (in 1991/2 dollars). Estimates of future costs depend crucially upon whether production will be coming from:

- deeper horizons of existing fields;
- satellites of existing fields;
- new grassroots projects (such as the Yamal Peninsula or the Barents Sea).

And whether transmission will be through:

- existing pipelines needing refurbishment;
- new pipelines needing to be built in complete (and possibly multiple) strings from field to export border;
- new lines which can be built in modular form (i.e. in stages) from field to border.

Each of the production variants (with the exception of Barents Sea gas) can be matched with a different transmission variant, giving a large number of potential cost permutations. For gas from the Yamal Peninsula fields arriving through new dedicated pipelines, two estimates are available:

- a delivered cost of (1994) $2.00–2.20/mmbtu at the Russian border (giving a price of $3.60 at the German border);
- a transport-only cost (including transit fees) of $2.50–2.75/mmbtu for Yamal gas delivered to Poland and Ukraine, which would give a price of around (1994) $4.00/mmbtu.

However, the problem with any reference to 'Yamal gas' is to know what is being referred to in terms of a transportation system and the origins of the gas passing through it. Map 1.3 identifies four stages of the Yamal pipeline project, each of which has a different cost estimate attached to it and no combination of which needs to be developed simultaneously. Only at the final (extremely costly) stage 4 are the Yamal fields linked into the system. Within each of stages 1–3, there is the option to build one pipeline or multiple pipelines and to add compression immediately or in stages. Hence the development permutations for Yamal gas are almost endless (see Table A.5 in Appendix to this chapter).

Map 1.3 The Yamal pipeline: stages of development

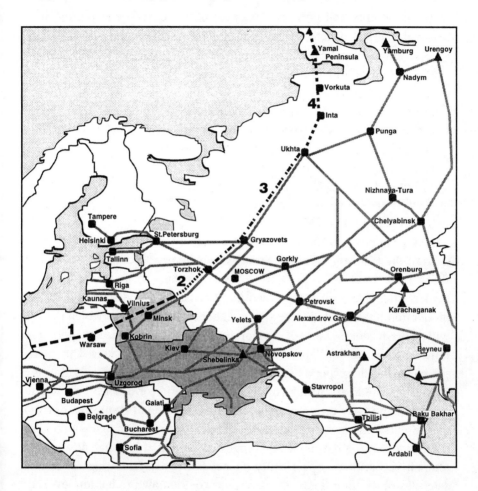

Source: Gas Strategies, London, 1995.

1.5 Production projections

In the economic and political upheaval of the post-Soviet period, it is reasonable to ask whether there is any point in attempting projections of Russian production beyond the current year (or at the most the next three years). Yet the culture of plans – annual, five-year and 'longer-term' – is so deeply rooted in both the Russian gas industry and long-term Western observers that the former continue to issue, and the latter to crave, numerical forecasts, irrespective of whether any credence can or should be attached to them. Table 1.4 shows a selection of forecasts and plans issued by VNIIGAZ (the main Russian gas research institute), Gazprom and the Ministry of Fuels and Energy during 1993–4. The 1995 estimates are already much too high, given the trends of the past year. It is generally believed that 1995 production will not exceed 600 BCM.[21]

Looking at the 5- to 15-year projections, the lowest estimate of production is from the Ministry's Energy Strategy, which sees figures as low as 715 BCM in 2000 and 785 in 2010; projections from the industry are significantly higher. This chapter has suggested that to make sense of future levels of production it is essential to be specific about the fields from which the gas will be produced; the costs of developing those fields; and the transportation infrastructure which will be necessary to bring the gas to markets. The reason for labouring these points is that, although there are elements within the Russian gas industry which continue to believe in (or at least continue to assert) the primacy of gas development from the Yamal Peninsula, or from the Barents Sea, these are the supergiant prestige projects characteristic of the Soviet era. What has been suggested here is that much lower-cost alternative options – specifically deeper horizons of existing fields and satellites of existing fields – exist which would allow production levels to be continued at 1990 levels at least up to 2010. But although economic arguments seem to go against developing the Yamal fields, political preferences may dictate otherwise. Within Gazprom there is a considerable momentum behind the upstream development of the project which would provide employment for a large part of the company's labour force. The

---

[21] 'Gas Output Forecast to Drop in 1995', *BBC Summary of World Broadcasts*, Part 1, SUW/ 0374 WD/1, 10 March 1995.

Table 1.4 Russian gas production projections, 1995–2010 (BCM)

	1995	1997	2000	2005	2010
VNIIGAZ					
Total production	680	705	735–55	785–820	820–60
of which					
W. Siberia	593	640	642–658	691–722	723–60
Yamal	—	—	0–10	20–45	93–135
Gazprom					
mid-1994	680		735–55		900–1000
late 1993	690–5		800–835		900–80
Ministry of Fuels					
and Energy	650–5		715–30		785–820

Sources: VNIIGAZ data cited in Cedigaz, *Natural Gas in the World, 1994 Survey*, table 40, p. 96. Gazprom forecasts from R.I. Vyakhirev, 'Nastoyashchie i budushchie postavki gaza Rossiyey v tsentralnuyu ir vostochnuyu Yevropu', paper presented to the World Gas Conference, Milan, June 1994; Rem Vyakhirev, 'Gazprom's European and Worldwide Marketing Plans', paper presented to the Financial Times Conference, Vienna, December 1993; B.V. Budzulyak, 'Upgrading and Retrofitting: Future Developments', in *Gas in the former Soviet Union, Petroleum Economist* Special Report, September 1993. Ministry of Fuels and Energy forecasts from *Russian Energy Strategy*, 1994.

unemployment consequences of failing to develop the Yamal fields would be distressing for Gazprom's management. Likewise in the Yamal region, concern about environmental disruption is balanced by anticipation of both employment for local people and revenues for local political authorities, both of which would be disappointed by failure to commence development.

Nevertheless, in terms of production projections, possibly the most interesting statistic in Table 1.4 is that the west Siberian fields outside the Yamal Peninsula could be producing well over 700 BCM in 2010. Whether anything approaching this level of production, will be required over the next 15 years, let alone the opening up of massive additional resources from the Yamal Peninsula and the Barents Sea, depends on the evolution of domestic and foreign markets for Russian gas.

Appendix: Cost estimates for production and transmission of Russian gas

Assembled in this Appendix are estimates from a number of Western and Russian sources, looking at the problem from a number of different perspectives. Much is unclear in the estimates, in which exchange rate calculations and assumed rates of return are the most important omissions. Tables A.1 and A.2 both give estimates of production and transportation costs.

Production costs

It is interesting that both sources give very low figures for upstream costs; although the IEA estimates in Table A.1 are lower by a factor of 4 than those in Table A.2, this may be accounted for by the fact that Birman is looking at the former Soviet Union as a whole, not simply at Russian gas fields. Also interesting is that in the IEA figures it is transportation which constitutes by far the largest component of delivered costs – which is probably the way most Western analysts would think – whereas for Birman production is the major cost. This may be explained by the fact that Birman's calculation assumes that most of the trunk pipelines will be fully depreciated by 1995 and that very little additional capacity will be required, at least to meet the production targets in Variant A.

Perhaps the most interesting feature of the Birman calculations is not the absolute but the relative increases in production costs, with a sevenfold increase between 1987 and 1992, a doubling in the three following years, but then an increase of only 20–50% for the low and medium production variants (A and B). It is only if the high production variant (C) is chosen that another step change in production costs is likely to occur. Dienes quotes modellers at the Institute for the Study of Complex Energy Problems as suggesting that production increases in the range of 1100–1200 BCM per year yield no net benefits because of the energy required to produce and transport the gas.[1]

---

[1] Leslie Dienes, Istvan Dobozi and Marian Radetski, *Energy and Economic Reform in the Former Soviet Union* (New York: St Martins Press, 1994), p. 79.

Table A.1 Production and transport cost elements in 1993 Russian gas prices to domestic and export markets (US cents/mmbtu)

	Domestic	FSU	Europe
Exploration and production costs	7.3	7.3	7.3
Exploration	1.8	1.8	1.8
Operating costs	3.6	3.6	3.6
Depreciation	1.8	1.8	1.8
Transportation costs	70.7	84.2	84.2
Operating costs	3.4	4.3	4.3
Depreciation	67.3	79.9	79.9
Total costs	78.0	91.5	91.5

Source: International Energy Agency, *Russian Energy Prices, Taxes and Costs* (Paris: OECD/ IEA, 1994), Tables 8a, 8b, 8c, pp. 56–8.

Transmission costs

Transmission costs have been approached in two different ways by different analysts: Tables A.1 and A.2 present very different totals, but it is the high depreciation figures in the IEA estimates that catch the eye. Given that these costs relate to existing systems which were given to Gazprom by the Russian government as part of a 'dowry' at the time of privatization, without any apparent debt liability, it is not certain that such high depreciation figures are appropriate. Despite this caveat, even the highest estimates suggest costs at the city gate, or export border, of around $1 per mmbtu (in 1991/2 dollars).

New production from existing fields

Tables A.1 and A.2 are not specific in terms of attributing costs to individual fields. However, Table A.3 attempts a ranking of such costs (albeit in relative terms) across a variety of sources including deeper horizons at existing fields, new satellites, Barents Sea fields, and onshore and offshore Yamal Peninsula fields.

It is a little difficult to understand how Grace's transportation figures are derived and therefore how he arrives at the total cost. It is important to distinguish between the two different elements of the calculation because, if

Table A.2 Projected FSU natural gas production and transport costs to borders or major consumption centres

	1987	1992	1995	2000 A	2000 B	2000 C
Production (BCM)	687	784	744	790	852	1000
Costs: (constant 1990 US cents/mmbtu)						
Production	4	28	63	75	95	285
Transportation	5	9	13	20	25	50
Total costs	9	37	76	95	120	335

Source: Birman, cited in Leslie Dienes, Istvan Dobozi and Marian Radetski, *Energy and Economic Reform in the Former Soviet Union* (New York: St Martins Press, 1994), Table 2.8, p. 77.

the figures in Table A.1 are correct, even if production costs at new fields are very much greater than at Urengoy, this would not necessarily raise the delivered cost of the gas significantly. Interestingly, Grace believes that the cost of existing production at Yamburg is double that of Urengoy, but that the cost of production in deeper horizons at Urengoy is substantially greater (25%) than that of deeper horizons at Yamburg (12.5%). The derivation of the 50–55% increase in transportation costs for gas from deeper horizons may be an allowance for separation and transportation of condensates.

Production and transportation costs from new fields
Nadym-Pur-Taz region   Grace (Table A.3) believes the cost of production from the Zapolyarnoye fields to be 250% higher than that of existing Urengoy gas, but he expects that Zapolyarnoye costs will be similar to costs from (Urengoy and Yamburg) satellite fields. This conflicts with the view expressed by Dienes that neither Russian nor Western technology is yet equipped to take on the Zapolyarnoye field.[2] Table A.3 also seems to indicate that transportation costs will be 25% above those of current Urengoy production; this seems a large increase given the relatively short distance from these fields to existing infrastructure.

---

[2] Ibid., p. 58.

Table A.3 Costs of new production relative to existing Urengoy gas

	Production	Transportation	Total
Urengoy I[a]	1.0	—	1.0
Urengoy II[b]	1.25	0.55	1.80
Yamburg I[a]	2.0	—	2.0
Yamburg II[b]	2.25	0.50	2.75
Zapolyarnoye	2.50	0.25	2.75
Urengoy satellites[c]	2.50	0.25	2.75
Shtokmanovskoye	4.0	1.0	5.0
Yamal onshore	5.0	2.0	7.0
Yamal offshore[d]	6.0	0.5	6.5

[a] Urengoy I and Yamburg I are costs of shallow (Cenomanian) gas in those fields.
[b] Urengoy II and Yamburg II are costs of gas in deeper (Neocomian) horizons in those fields.
[c] Gubkinskoye, South Russkoye, East and West Tarkosalinskoye, Komsomolskoye and Yubilyeynoye.
[d] Assumes that the onshore fields will be developed first, and all common costs of Yamal production will be loaded on to the onshore fields.
Source: John D. Grace, 'Cost Russia's Biggest Challenge in Maintaining Gas Supplies', *Oil and Gas Journal*, 13 February, 1995, pp. 79–81.

Yamal Peninsula   Grace (Table A.3) divides the Yamal Peninsula into onshore (Bovanenko/Kharasavey/Kruzenshtern) and offshore (Leningradskoye and Russanovskoye). He sees onshore production being five times (and offshore six times) the cost of existing Urengoy gas, but again the transportation costs are somewhat confusing. He may be suggesting that transportation costs of Yamal gas will be three times the transportation cost of (i.e. 200% more than) current transportation from Urengoy.

The figures in Tables A.1 and A.2 do not take into account the additional costs of a pipeline originating in the Yamal Peninsula, which Dienes notes would be 'far more expensive', because of: the crossing of the Gulf of Baidarat; the need to create a corridor in virgin territory (at least as far as Ukhta); and the much longer span of continuous permafrost to be covered. Dienes quotes Russian sources suggesting that capital outlays per unit for Yamal fields would be three to four times higher than average comparable figures for Urengoy and Yamburg, with a production level of 160 BCM at Bovanenko (plus the pipelines to transport the gas to markets) requiring a

Table A.4 Approximate costs for different components of a three-pipeline Yamal Project

Annual volumes: 83 BCM production
                 60 BCM exports at the Russian border

Stage 4: $20–22bn
Field development; transmission to the Gulf of Baidarat (three 56 inch pipelines plus 2 compressor stations); crossing the Gulf (four 48 inch pipelines) – $10–12bn

Gulf of Baidarat to Ukhta (three 56 inch lines, five compressor stations) – $10bn

Stage 3: $12bn
Ukhta–Torzhok (three 56 inch lines, eight compressor stations) – $12bn

Stage 2: $2bn
Torzhok–Belarus border (two 56 inch lines, 4 compressor stations) – $2bn

Stage 1: $4.7bn
Belarus–Polish border (two 56 inch lines, 5 compressor stations) – $2.2bn

Poland–German border (two 56 inch lines, 5 compressor stations) – $2.5bn

Grand total $38.7–40.7bn

Source: *Proekt Yamal-Evropa* (Gazprom, 1994).

capital investment of $56 bn (1990).[3] By contrast, Gazprom (Table A.4) arrives at a figure of around $40 bn (1994) for a three-pipeline project, of which $10–12 bn is specifically for work on the Peninsula.

These capital costs are not necessarily inconsistent with the estimates in Table A.5 which suggest a total transport cost (including transit fees) of $2.50–2.75/mmbtu for Yamal gas delivered to Poland and Ukraine. However, the 'normal profit' allowed for in these estimates is unlikely to approach the 15% 'hurdle rate' generally used by international companies which, with the addition of $0.5–1.00/mmbtu for production (capital and operating expenses), would bring the costs of Yamal gas to the generally accepted Western estimate of $4.00/mmbtu. By contrast, the Chairman of Wintershall has been reported as saying that Yamal gas would be priced at the German border at $3.60/mmbtu and that the delivered cost would be $2.00–2.20/mmbtu.[4]

[3] Ibid., p. 76.
[4] 'Yamal Gas $130/mcm at Border', *International Gas Report*, 25 November, 1994, p. 9.

Table A.5 Transport costs of gas from the Yamal Peninsula to export borders (US\$/mmbtu)

	Belarus/Poland	Ukraine
Distance (km)	6000	6500
Third country crossing (km)	1600	2100
Technical cost[a]	1.8–2.0	2.0–2.2
Transit fees[b]		
A	0.3	0.45
B	0.5	0.60
Total transport cost[b]		
A	2.20	2.50
B	2.40	2.75

[a] Total technical cost per unit of transportation including 'normal profit'.
[b] In variant A, third countries take 5% of the gas which flows through their territories; in variant B third countries charge 3 cents per mmbtu per 100 km.
Source: Observatoire méditerranéen de l'Energie, cited in *Natural Gas Transportation: Organisation and Regulation* (Paris: OECD/IEA, 1994), Table 3, p. 141.

The real problem with all of these estimates is to know exactly what is being referred to. The 'Yamal project' now encompasses such a wide variety of production and transmission options that it is no longer possible to distinguish how many lines are being discussed, running from which fields to which destinations. Because of this, Table A.4 is intended to allow readers to make their own approximate estimates based on the number of pipelines and the stages which may be built. Table A.4 is based on a 'three-pipeline' Yamal option and an approximate figure of \$40 bn for the total capital cost. If the cost estimates presented in this table can be accepted as approximately correct then it is clear that a 'Yamal project' which involved only a single 56 inch pipeline from the Russian/Belarus border to the Polish/German border would have a cost of around \$2.5 bn, while a 'Yamal project' which involved two 56 inch lines built from Torzhok to the German border (Stages 1 and 2) would involve a cost of around \$6.7 bn. It is only when field development and the major lines through Russia commence (Stages 3 and 4) that investment costs become much larger.

These figures are somewhat higher than Dienes's earlier attempts to estimate the costs of new lines from existing fields. Working from a mixture of Russian and Western data, Table A.6 provides the components of two pipelines: a 3000 km pipeline from the western slopes of the Urals to the Slovak border requiring an investment of $3.75–3.95 bn, and a 4000 km pipeline from the Yamburg area to the Slovak border, requiring $5.23–5.43 bn.

It is worth emphasizing a number of points about the approximate nature of these figures. First, they are based on historical data where costs have been scaled up to 1990s levels. Second, they are for single strings of pipeline which may not take advantage of the economies of scale which are possible for multiple pipelines. Third, the distance between Urengoy and the Slovak border is generally stated as 4500 km, so that a Yamburg to Slovakia line might approach 4600 km rather than the figure of 4000 km which is used here. A final point to note about these ranges is that they seem significantly below the figures which were quoted for the original Urengoy pipeline to Slovakia, which was generally believed to cost 7.6 bn roubles (1983) or around $10 bn (1982).[5] Built five years later, the Yamburg pipeline to the Slovak border was reckoned to have cost 10 bn roubles (1988).[6]

Calculated on the basis of the figures in Table A.6, the cost of a single 1000 km line from the Torzhok compressor station on the Northern Lights pipeline through Belarus to the Polish border would be around $1.3–1.5 bn; this compares with a cost of around $2.1 bn derived from the figures in Table A.4. The cost of a 2500 km line from Ukhta (where the lines from the Yamal Peninsula would meet the existing Siberian corridor) to the Polish border would be around $3.25–3.75 bn on the basis of figures in Table A.6, compared with $6.1 bn calculated from Table A.4. Some part of the difference can be accounted for by inflation since Dienes made his estimates.

Production and transportation costs for the Barents Sea field
In contrast to the complex appraisals which have been completed for the Yamal fields and pipelines, much less detailed work has been carried out on a possible Barents Sea project. If the Russian estimate that $6–7 bn is needed to develop the Shtokmanovskoye field is directionally correct, the shorter

---

[5] Marie Françoise Chabrelie, *European Natural Gas Trade by Pipelines* (Ruiel Malmaison: CEDIGAZ, July 1993), p. 13.
[6] Ibid., p.16.

Table A.6 Approximate costs of new Russian gas pipeline construction, 1992 ($ bn)

	Western Urals–Slovakia	Yamburg–Slovakia
Distance (km)	3000	4000
Imported equipment:		
Pipe[a]	1.3	1.8
Compressors[b]	1.3–1.5	1.7–1.9
Other equipment	0.75	1.2[c]
Domestic labour[d]	0.36	0.53
Total	3.75–3.95	5.23–5.43

[a] Imported Japanese pipe at $730/ton plus freight charges of $75/ton (1992 figures).
[b] Lower figures are for a throughput of 28 BCM per year, higher figures for 36 BCM per year. Source suggests that an upper limit of 30–33 BCM per year may be more realistic.
[c] Source suggests a figure of 'well over $1 bn'.
[d] Converted at the 1990 exchange rate of R3.2=$1 from R1.15 bn for the shorter line and R1.7 bn for the longer. Cost escalation from rouble investment for Urengoy of 15% applied to 3000 km line and 30% to 4000 km line.
Source: Dienes et al., *Energy and Economic Reform*, Appendix 2.3, pp. 118–19.

transportation distance (compared with the Yamal alternative) would bring gas to Europe at lower cost; it would also give the Russians an alternative transportation route which might completely avoid the former Soviet Union.[7] Cost estimates from Western sources are somewhat higher at around $10 bn for a single pipeline carrying 30 BCM per year. These sources suggest that $3.00/mmbtu (1993) could be a realistic price for Barents Sea gas delivered to Europe. But this would be for a single pipeline development without the rather ambitious LNG and methanol facilities foreseen in Russian sources.[8] The estimates in Table A.3 support the contention from commercial companies that if the Barents Sea is compared with a grassroots Yamal project (as opposed to an incremental Yamal project), Shtokmanovskoye would be the more cost-effective option.

[7] 'Rosshelf and the Barents Sea', *Eastern Bloc Energy*, August 1994, pp. 4–5.
[8] E. P. Velikhov, 'Development of the Shtokmanovskoye Gas/Condensates Field', in *Gas in the Former Soviet Union*.

Chapter 2

The Demand Side: Sectors, Regions, Prices and Payments

Historically it was always immensely difficult to obtain detailed data on Soviet demand for energy, including natural gas. Estimates were based on 'apparent' consumption – production minus net exports. Even this required certain assumptions to be made about own-use, storage and losses. Taking these caveats into account, natural gas demand in the former Soviet Union increased very strongly, by more than 5% annually, throughout the final two decades of the Union's existence.[1] Gas was substituted for oil in industry, and in power generation, where it also had to compensate for the much less than intended availability of coal-fired and nuclear power plant.

In the post-Soviet period availability of statistics for the Russian gas industry has improved greatly, but clarity is still lacking in important aspects of the demand picture. This chapter seeks to deal in some detail with the component parts of Russian gas demand, attempting to look at individual sectors and regions, as well as issues of prices and payments. This information is used to make projections for the period up to 2000.

2.1 Gas demand

In many respects the 'apparent consumption' approach, which treated demand as a residual factor in the gas balance, was not necessarily incorrect during the Soviet era. All of the official statistics concentrated on production plans and achievements. Towards the end of the period, lip-service was paid to the notion of energy conservation and efficiency; but even if the intention was to take these concepts seriously, the absence of cost-based pricing and consequent lack of incentives to save energy made it almost impossible to put such ideas into practice.

---

[1] Leslie Dienes, Istvan Dobozi and Marian Radetzki, *Energy and Economic Reform in the Former Soviet Union* (New York: St Martins Press, 1994), pp. 122–3.

Table 2.1 Russian natural gas balance, 1990–1993 (BCM)

	1990	1991	1992	1993
Production	640.2	643.0	640.4	617.6
Storage (net)	2.9	2.0	2.0	12.8
withdrawals	24.9	29.1	30.4	28.3
additions	27.8	31.1	32.4	41.1
Total exports:	202.1	195.2	189.1	179.5
to non-CIS[a]	110.1	105.2	99.1	100.9
to CIS	92.0	90.0	90.0	78.6
Pipeline consumption	56.7	59.7	59.4	42.4[b]
Internal consumption	378.5	386.1	389.9	382.9
including:				
power	179.0	179.5	176.8	163.9
metallurgy	37.2	36.5	32.8	34.4
chemicals	31.1	30.5	26.2	27.0
cement and building materials	21.1	20.9	20.8	19.8
Other[c]	110.0	118.7	133.3	137.8

[a] These are in fact former Soviet and CIS exports to Europe, including the 'quota' from Turkmenistan.
[b] Estimate.
[c] Includes other branches of industry and municipal consumers.
Source: data from VNIIGAZ reproduced in Sylvie Cornot-Gandolphe,
Le gas naturel dans le monde (Rueil Malmaison: Cedigaz, 1994 ed), table 39, p. 95.

Table 2.1 shows the standard Russian statistics on gas demand prepared by VNIIGAZ. While the figures are robust in some respects, they also contain some inconsistencies which are investigated below. (The data on trade are incomplete in that they do not include imports; we return to this issue in Chapter 3.) However, these data are the most complete and consistent demand statistics in the public domain, and have been updated for each of the past three years. It is worth focusing on some elements of the balance in more detail.

Net storage additions
The figures show both additions to and withdrawals from storage. The comparatively large net addition of 12.8 BCM in 1993 reinforces the Gazprom

assertion that production is being shut in because of lack of markets, both domestic and export. There are indications that similar net storage additions of 9–10 BCM were recorded for 1994.[2]

Pipeline fuel and system losses

This is always a contentious issue in any consideration of Russian gas demand. It is virtually impossible to distinguish between the use of fuel for compression and leakages from the system. Gazprom has suggested that around 10% of total production is used as fuel for the transmission system, but the VNIIGAZ figures in Table 2.1 are somewhat lower; in particular the 1993 figure for pipeline consumption is not explicable.

In the category of 'losses', Gazprom has suggested a figure of 0.8–1.2% of total production.[3] This may seem too low, but as long as it is taken to apply only to the high-pressure transmission system it may be credible. Larger losses may occur in the distribution system, over which Gazprom has no control and where even less information is available. Lack of metering in the entire residential sector means that it would not be possible to put a figure on these losses and they would simply be recorded as 'demand'. A weakness of the figures in Table 2.1 is that they appear to take no account of losses.

Internal consumption

The figures under this heading are intended to denote volumes actually delivered to consumers (with the caveat noted above regarding the treatment of losses). The trend in the VNIIGAZ figures for 1990–92 is upwards but this is based on a strong increase in the 'other industry' category. The clear downward drift in demand for all energy-intensive sectors of industry is closer to reality; the 1993 increase in demand for metallurgy and chemicals

---

[2] Interfax news agency reported that during the period 1992–4 a 'surplus' of 24 BCM had been put into storage: *BBC Summary of World Broadcasts* (henceforth *SWB*), Part 1, SWB/ 0376 WD/1, 24 March, 1995.

[3] A. M. Boiko, 'Refurbishing the Gas Transportation System', paper presented to the conference on, 'Natural Gas: Trade and Investment Opportunities in Russia and the CIS', Royal Institute of International Affairs, London, 13–14 October, 1994 (henceforth RIIA Conference).

is not explicable and probably not correct.[4] Possibly the most significant pointer to the future is the 7.3% (12.9 BCM) drop in power generation demand for 1993.

The figures in Table 2.2 show a more detailed picture of sectoral demand but the methodology by which they have been constructed is different from, and cannot easily be compared with, Table 2.1. The figures for 1990 differ from those of 1992 and 1993 in terms of end-uses categories, and comparisons are therefore not exact between these data. However, these more detailed figures repeat the picture shown in Table 2.1 with a continuing steady decline in power generation and a corresponding, but smaller, decline in industrial production. The decline in industry fails to reflect the huge drop in industrial production during this period and (as noted above) the increase in metallurgy demand is very difficult to account for. These figures also come with omissions and contradictions, and have been interpreted by the author.

The power generation sector
The single most important area of gas demand is power generation. Even with the decline of gas use in power generation to 165 BCM in 1993, this sector still accounted for nearly 43% of Russian internal gas demand. For that reason, in any attempt to project gas demand into the future specific attention needs to be devoted to power generation.

Table 2.3 shows Russian power generation capacity by region and fuel type in 1991.[5] It shows that of the 174.77 GW of operational capacity in 1991, gas-fired stations accounted for 37% and dual-fired gas/fuel oil stations for another 2%.[6] Table 2.4 concentrates on the gas and dual-fired stations by region, showing that outside Siberia the vast majority of capacity is

---

[4] Production in the metallurgical industries fell by 10–20% in 1993; production of chemical products fell by 20–30%: *Russian Economic Trends*, vol. 2, no. 4, 1993, table 35, p. 449. It seems highly unlikely that interfuel substitution in these industries caused gas demand to rise significantly in 1993. If gas demand did indeed rise, it can only have been due to deliveries for which no payment was made.

[5] The sources for these figures and those in Tables 2.3 and 2.4 are drawn from a power generation database to which this author has had access which gives detailed statistics for every power station in Russia. Much of the detail can be gained from International Energy Agency, *Electricity in European Economies in Transition* (Paris: OECD, 1994), pp. 212–23.

[6] The theoretical 'nameplate' capacity of power stations in 1991 was over 190 GW; the figure of 174.77 GW represents operational capacity in that year.

Table 2.2 Russian natural gas demand by sector, 1990–1994 (BCM)

	1990	1992	1993	1994[c]
Power generation	179.0	174.0	165.4	158
Industry	182.7	150.4	149.1	145
oil/gas	24.8	12.1	9.8	
RAO Gazprom own-use		9.6	9.3	
construction	21.2			
RosCement		6.8	6.4	
metal working and machine building	31.3			
automotive/agricultural machinery	5.9	5.8		
non-ferrous metallurgy	8.4	} 30.8	34.2	
ferrous metallurgy	28.8			
chemicals (total)	31.1			
agro-chemical/fertilizers		18.4	17.6	
petrochemicals		8.5	7.6	
agro-industrial complex		17.5	17.6	
defence		6.3		
other[a]	20.1	34.5		
Other[b]	17.0	17.3	13.4	
Residential and commercial	142.3	53.6	55.0	57
Total	404.0	395.3	382.9	360

[a] Light industry, food, others.
[b] For 1990: transportation, construction, agriculture; for other years: all sectors apart from those itemized.
[c] Estimate.
Note: 1990 figures use Soviet end-user categories. 1992 and 1993 figures use different methodology and are not strictly comparable with the 1990 figures.
Sources: PROMGAZ and GAZPROM data interpreted by the author. See also: 'Gazprom Study Cracks Open Secrets of the Gas Giant', *World Gas Intelligence*, 27 January, 1995, p. 5.

concentrated in four regions: Urals, North Caucasus, Middle Volga and Centre, with the last accounting for more than the other three combined. The final column in the table shows 'imputed' gas demand, which is derived by taking the total gas demand figure from Table 2.1 and apportioning this figure among regions based on generating capacity in each region.[7] Table 2.4 shows

---

[7] While this would not be as accurate as calculating the heat rate and load factor of each station, it probably represents a reasonable generalization for regional demand.

Table 2.3 Russian power generation capacity by type of fuel and region, 1991 (%)

	Oil	Gas	Gas/Fuel Oil	Coal	Hydro	Nuclear	PSO[a]
Centre	2	54		15	8	20	1
Middle Volga	13	40	5	4	23	15	
North Caucasus	2	56		23	19		
North West	20	4	9	9	18	38	4
Urals	4	36	6	45	5	2	2
West Siberia		99					1
East Siberia/Far East	2	1		46	49		2
Russia total	5	37	2	25	19	11	1
in GW[b]	8.48	63.95	4.23	43.59	33.69	19.11	1.70

[a] Peat, shale and other.
[b] Operational capacity as of 1991; total = 174.77 (allowing for rounding).
Source: see note 5.

the importance of the Centre, Urals and Middle Volga in non-Siberian gas-fired power demand.

2.2 Prices and payments

Price issues

Any serious discussion on Russian gas prices needs to take account of:

- the level of gas prices to different Russian consumers;
- how these are, and should be, calculated;
- the extent to which consumers are paying, and can be made to pay, those prices;
- the announced plans for price reform, whether these are likely to be implemented and on what timetable.

These issues are worthy of a separate chapter in their own right; here we can deal only with a small number of relatively recent issues in the certain knowledge that the situation is likely to change radically.[8]

[8] For background see International Energy Agency, *Russian Energy Prices, Taxes and Costs* (Paris: OECD/IEA, 1994), ch. 4 and table 15, p. 91.

Table 2.4 Gas-fired power generation capacity in Russia, 1991

| | Natural gas | | Gas/Fuel oil | | Imputed[a] gas demand |
Region	MW	% of total	MW	% of total	BCM
Centre	28,738	44.9	287	6.8	77.4
Middle Volga	8,075	12.6	1,007	23.8	23.6
North Caucasus	5,680	8.9			15.2
North West	573	0.8	1,212	28.6	4.0
Urals	10,608	16.6	1,726	40.8	31.8
West Siberia	9,901	15.5			26.5
East Siberia/Far East	412	0.6			1.1
Russia total	63,951	100	4,232	100	179.5

[a] Assuming dual-fired stations use 75:25 gas and fuel oil
Source: see note 5.

From the middle of 1993, a base was established for industrial gas prices which was indexed to inflation. Prices are increased monthly according to this index.[9] Industrial gas has been priced uniformly throughout Russia, with no allowance made for differential transportation costs or for differing values of gas to end-users. In November 1994 the wholesale industrial gas price stood at 55,020 roubles per thousand cubic meters (R/mcm) and the retail price at 62,679 R/mcm.

Residential gas prices do not appear to be set in the same way as industrial prices. They are raised periodically, but much less frequently. In November 1994 wholesale residential gas prices stood at 600 R/mcm and retail prices at 2000 R/mcm. However, since residential consumers have no meters on their premises they are charged according to a per person, or per square metre, 'norm' for consumption, which means that the gas bill of the average family at the end of 1994 was around 50–60 roubles per month.[10]

---

[9] In fact, the indexation formula appears to be a 'horse-trade' between Gazprom and the government with inflation being the main, but not the only, factor. Prices are raised not monthly but irregularly, depending on how long the Government Price Committee takes to come to a decision.

[10] At an exchange rate of $1 = R3000, at a time when a bottle of mineral water cost around R1000 in Moscow. The final decision on residential prices appears to be made in the Prime Minister's Office.

It is clear that much remains to be done in terms of raising and revising Russian gas prices, and there are many indications that such measures are being actively contemplated and (possibly) implemented. During 1994 no progress was made in raising industrial gas prices in real terms.[11] On 1 March 1995 a new regulation was issued increasing the excise tax to 25% of the total price for all consumers, and raising the price to households to 20,000 R/mcm (i.e. tenfold).[12] In addition, there were reports that industrial gas prices had also been raised significantly and would be differentiated by region.[13]

Whether these measures will be implemented, and whether they will have any effect, will depend on the enforcement of payment (discussed below). The most positive aspect is that Gazprom believed that – even at 1994 prices – it was profitable to serve industrial consumers. But at 1994 prices, no new gas project – certainly not any of the massive capital-intensive projects discussed in Chapter 1 – could sell gas profitably in the Russian market. The general problems of the 1994 pricing structure were three in number:

First, the structure of prices was the opposite to that which the cost structure of the industry would suggest, with industrial consumers expected to pay a unit price 30 times greater than residential consumers, although the cost of serving the latter is clearly much greater. (Following the March 1995 price increases, the differential between industrial and residential prices was reduced to a factor of around six.) This price structure gave strong incentives to sign up industrial gas users and avoid residential customers as simply a profitless nuisance. Anecdotal evidence suggests that some part of the sharp rise in residential/district heating demand shown in Table 2.2 may be due to consumers 'reclassifying' themselves from the industrial sector in order to pay lower prices.

Second, residential customers were receiving gas virtually as a free good. Their payments – when they paid – barely covered the cost of bill-collection. This is important because, as we shall see below, significant increases in

---

[11] 'Zdes ozhidaetsa stabilnost', *Ekonomika i Zhizn*, no. 5, February 1995, p. 4.

[12] *On Regulating State Natural Gas Prices* (Moscow: Government of the Russian Federation, Regulation no. 208, 28 February, 1995).

[13] On 1 March, 1995 the retail industrial gas price was increased to R137,809/mcm. *Gas Matters*, April 1995, p. 11.

demand are expected in this sector. In such a situation, any plans to install meters or connect new customers are out of the question.

Third, assuming that the gas is being paid for, industrial gas prices were probably not far from the level needed to promote conservation and efficiency measures. However, in terms of opportunity cost, the level of industrial prices in spring 1995 was around two-thirds of the European export (border) price.[14]

Payment issues

Gas demand figures for 1993 and particularly 1994 are significant in that they will have been affected not only by falling industrial production but also by the current curse of Russian industry; non-payment of bills. A 1994 estimate suggested that 60% of the decline in demand during the year was due to contraction of industry while 40% was due to non-payment. The non-payment problem affects the whole of the Russian market (as well as exports to former Soviet republics). According to Gazprom, during 1994, 53% of gas delivered to Russian consumers had not been paid for.[15] Extrapolating on an annual basis (using an estimated 1994 demand figure of 359 BCM), this would mean that more than 190 BCM of gas – nearly twice the volume of Russian exports to Europe – had not been paid for.

The solution to non-payment is not simple, since both politically and logistically it is not possible for Gazprom or the distribution companies to cut off a large proportion of their customers. In some circumstances it is impossible to isolate individual consumers for disconnection because of lack of infrastructure. This applies not only to individual residential apartments in blocks, but in some cities also to industrial consumers. A decree of May 1994 prohibits the disconnection of 'strategic' consumers on grounds of non-payment. This applies to power generation customers and the entire residential/commercial sector. In addition, regional political authorities exert

---

[14] An exact comparison between industrial gas prices and European border prices is complicated by assumptions about transportation tariffs. Converting the 1 March, 1995 industrial price into dollars at the current exchange rate produces a figure of $39.20/mcm, compared with a European border price of around $63/mcm. However, some allowance needs to be made for additional transportation costs to take gas to the Russian border.

[15] 'Russian Gas Company Cuts Back on Drilling in Western Siberia', *SWB* SUW/0371 WD/ 5, 17 February 1995.

strong influence on gas suppliers not to cut off industries and factories which are major employers in their particular regions.

Despite these difficulties, it is known that Gazprom has begun to disconnect customers, even large power stations, for non-payment. One of the most important issues for the future will be the extent to which non-payment will constitute grounds for disconnection, and the length of time necessary to install the infrastructure to implement disconnection. Until that time, continued deliveries to customers – particularly enterprises – which either cannot or will not pay for gas will be dependent upon a combination of Gazprom and (central and regional) government discretion. The increase in government taxation of Gazprom sales to the Russian market in early 1995 may be an indication that a decision has been taken to allow Gazprom more freedom to disconnect. Anecdotal evidence is available that Gazprom is operating discounts on official selling prices for prompt payers.

For as long as non-payment is allowed to continue to any significant extent, all calculations of 'demand' are fatally flawed. 'Demand' is a concept which presupposes that those to whom gas is being delivered are paying an agreed price, and that if they fail to pay they will not be supplied. In the present situation, Russian gas demand is what Gazprom and the government decide it should be. Until it becomes clear how many consumers cannot pay (i.e. are bankrupt), as opposed to how many are finding it convenient not to pay, thereby using gas suppliers as a cheap source of credit (which is very profitable at high rates of inflation), it will be impossible to arrive at levels of real demand at current prices. Moreover, it may be counterproductive to continue to raise the 'official' selling prices for gas because – although this has a superficial attraction for market economists and international lending agencies – it may simply deter those enterprises which are paying their bills from continuing to do so.

2.3 Projecting future gas demand

Macroeconomic projections
Broad-brush statistics suggest that Russian industrial production fell by more than 50% in the period 1991–4. Industrial production in 1994 was around

21% below the figure for the previous year.[16] In many respects, these figures may be less dire than they first appear. Some part of this decline can be accounted for by the contraction of the military sector, with the civilian economy being far less badly affected. Another part of the fall can be ascribed to under-reporting of output by enterprises in an attempt to avoid taxes. A third factor is that production from small, recently founded enterprises does not appear in the statistics. There is ample anecdotal evidence of rising consumption trends and living standards in some regions and sectors of the population. For our purposes, the figures for decline in production for individual industrial sectors, expressed in physical units, are more important than the (less easily defined and more problematic) GDP figures, because they should provide a better indication of expected movements in demand for energy – specifically, for gas and electricity.

Studies of Russian energy demand typically use a projection of GDP growth rates combined with an energy/GDP ratio in order to arrive at demand figures. Some studies also apply an economic and political reform label to a time horizon such as: 'trends continued', 'rapid reform', 'failed reform', etc. There is a tendency to use international comparisons to demonstrate the very high energy intensity of the Russian economy, and hence the potential for large energy savings.[17] Two principal problems of the 'top-down' approach in the current Russian situation are the weakness of the GDP figures and the lack of any specific guidance as to how far and how fast the Russian economy might approach the energy intensity of OECD countries. What can be said with certainty is that the potential for very large savings not only exists but, despite being acknowledged over many years by Soviet and Russian leaders, remains relatively untapped. However, noting this does not help us to understand how rapidly this process could move forward.

Here we attempt a more modest, but still relatively complex examination of the sectoral and regional components of Russian gas demand, in order to

---

[16] Goskomstat data recorded in 'Russia's Economic Performance in 1994', *SWB* SUW/0368 WA/4, 27 January, 1995.

[17] See for example the comparisons in Dienes et al., *Energy and Economic Reform*, table 3.2, p. 131, and the conclusion (p. 135) that the theoretical savings potential of the former Soviet Union in 1990 was around 500 million tonnes of oil equivalent, compared with an actual demand three times that figure.

Figure 2.1

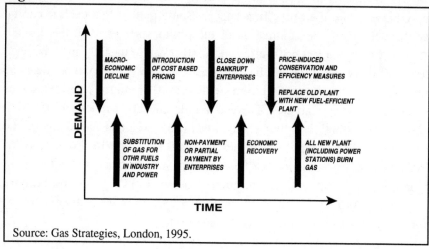

Source: Gas Strategies, London, 1995.

try to make plausible assumptions about future trends. Quite aside from the political and economic uncertainty, projections of gas and electricity demand require an analyst to take a position on major macroenergy (and macro-economic) factors, the extent and timing of which can only be guessed. Figure 2.1 indicates these forces, showing their most likely sequence and how their effects might be offset over time. The figure incorporates the (questionable) assumption that Russia will continue, however slowly, on its current path to market-based reforms (especially cost-based pricing).

The first stage, on which the country is well advanced, is macroeconomic decline, but any decline in total energy demand may be offset by substitution of gas for other fuels: oil in industry, and coal and nuclear power in power generation. At the next stage, prices are raised to levels which begin to reflect costs, but enterprises fail to pay their bills (or do not pay the full amount of the new prices) and simply run up massive debts to the gas and electricity industries. At the next stage, many – or even most – of the loss-making enterprises are closed down through bankruptcy, but economic recovery in these and other sectors begins to push up energy demand. Finally, the full introduction of cost-based pricing leads to major conservation and efficiency measures along with the replacement of old equipment by new energy-efficient plant. However, although new plant will be much more efficient than that which it replaces, the vast majority of it will be gas-fired.

Table 2.5 Russian gas demand projections, 1995–2010 (BCM)

	1995	1997	2000	2005	2010
Power	221	235	244–248	275–285	288–297
Heat generation	80	78.2	81–83	83–86	85–90
Industry[a]	101	107.3	103–105	100–105	97–103
Raw material	20	21.5	23–24	24–26	26–29
Pipeline use	63	62	65–67	70–75	75–80
Total	485	504	516–527	552–577	571–599

[a] Metallurgy, machinery, building materials and other branches.
Source: data from VNIIGAZ, cited in Cedigaz, *Natural Gas in the World, 1994 Survey*, table 40, p. 96.

The period up to 2000

The most widely available Russian gas demand projections are shown in Table 2.5. These show demand rising slowly but steadily up to 2010, driven mainly by power generation. This appears to be a typical 'Soviet-type' projection in which figures move relentlessly upwards. It is presented here to indicate the general thinking about demand which, until relatively recently, appeared to be driven by the need to find a home for expected levels of production. These projections are reinforced by public statements from Gazprom continuing to suggest that demand in 2010 will be 30% above that of 1993.[18]

The likely decline in gas demand

For our purposes, the key outcome which is so difficult to project at this turning-point in the economic history of Russia is how the future industrial restructuring of the country will unfold in both a sectoral and a geographical perspective. Any forecast needs to concentrate on both

- sectoral projections, of which power generation and heavy industry are the most important;
- regional projections of the location of decline and the location of recovery in different industries, recognizing that new industry (even in the same sector) may not necessarily remain in the region in which the old plants were located.

[18] R. I. Vyakhirev, 'The Russian Gas Industry in a National and Global Context', paper presented to RIIA Conference.

Table 2.6 Russian Gas Demand Decline Scenarios 1990–2000 (BCM)

| | | | 'Forecast 2000 | | |
| | | | | Author | |
	1990	1994 (est)	Gazprom	% decline in sector	BCM[a]
Power generation	179.0	158.0	167	30	140
Industry	165.7	145.0	149	60	95
oil/gas	24.8		23	50	12
construction	21.2		18	60	11
machine-building and metal-working	31.3		26	60	17
non-ferrous	8.4		9	40	6
ferrous	28.8		25	70	14
chemicals	31.1		27	70	15
other[b]	20.1		21	0	20
Residential/district heating	42.3	57.0	65	+40	68
Other[c]	17.0		20	20	15
Total	404.0	360.0	405–410		318 (300–340)[d]

[a] Note that gas decline cannot simply be derived from sectoral decline because of fuel substitution effects.
[b] Light industry, food, others.
[c] Transportation, construction, agriculture.
[d] Likely range, see text.

Sectoral decline   Table 2.6 is an attempt to chart the potential decline in gas demand in industry and power generation up to 2000, showing scenarios calculated by Gazprom and by the author. The interesting feature of Gazprom's projection is that it breaks with tradition in showing total demand in 2000 at around the same level as in 1990 – and, crucially, it shows gas demand both for industry and for power generation as significantly lower. The only significant growth is in the residential and district heating sector; the 'transportation, construction and agriculture' figure shows some increase, while of the entire heavy industrial sector only non-ferrous metals shows some demand growth.

The author's scenario shows an estimate of the potential decline in each sector, followed by calculations of gas demand for those sectors corresponding to that decline. The decline in power generation is calculated on base-load stations only; no decline is expected from generating capacity using gas for peak load. Industrial sector decline is focused on heavy industry. Given the lack of available data, assumptions have been made about the extent of gas substitution in the different sectors.

Three important points need to be made about these estimates. First, the projected decline in each sector is in no way the maximum which could be expected. Given the decline which has already been experienced by the end of 1994, this author believes that the fall in demand could be significantly greater. Second, there is no simple correlation between the projected decline in gas demand and sectoral decline. Account has been taken of the likely fuel substitution effect, which means that in all cases (with the exception of the oil and gas sector), the decline in gas demand is significantly less than the decline in sectoral production. Third, although the single year of 2000 has been chosen, the figures are intended to reflect demand in the industrial sectors at the moment when economic recovery begins. That is to say, the demand projection reflects a judgement of industrial demand at the point where factories are paying cost-related prices with non-payment significantly lower than in 1994 – a point by which the major necessary plant closures have been achieved. No significant economic recovery or industrial growth (in these sectors) is assumed to have taken place, and no significant efficiency/ conservation/plant replacement measures to have been implemented. The problem is clearly that it is artificial to designate a single year as the point at which all industries will reach this same stage of development. Some industries will arrive at this point before 2000; some may take rather longer to get there. Therefore, the specific figure of 318 BCM for demand in 2000 is probably better seen as a range of 300–340 BCM reflecting the possibility that some sectors will have begun to recover strongly by 2000, while in others gas demand will continue to decline (although production may start to recover). Conservation and efficiency measures and replacement of plant will mean that even sectors which are growing strongly need not necessarily be experiencing significant increases in gas demand.

A further aspect of Table 2.6 worth noting is the significant increase in demand for the residential and district heating sector. This reflects a conviction that – despite the potential savings to be made in this sector from price-induced conservation through metering and thermostatic controls – the rapidly growing service sector (which would correspond to 'commercial demand' in OECD parlance) will give rise to major increases in gas consumption. Against this, there will come the combined shock of massively increased prices together with enforcement of payment (eventually). In the residential sector (especially district heating) there is likely to be a continued delay in implementing these measures for logistical and political reasons.

The general conclusion, therefore, is that Russian gas demand could be 64–104 BCM lower in 2000 than in 1990, and 20–6083 BCM lower than in 1994. However, it is worth noticing that these are total figures, arrived at after other sectors have been taken into account. The decline in power generation is estimated to be only 11 BCM compared with 1994 (40 BCM compared with 1990), with industrial gas demand declining by 55 BCM (70 BCM compared with 1990).

Regional decline   Table 2.7 attempts to identify the geographical location of possible demand decline more precisely. This table has been derived from three sources: the power generation database referred to in note 5 above; gas distribution figures which show consumption by oblast; and data on industrial production by oblast.[19] What is attempted in Table 2.7, using some of the same assumptions as in Table 2.6, is to derive figures for likely surplus gas demand by oblast, differentiating between power generation and industry. One difficulty in comparing Tables 2.7 and 2.6 stems from the fact that they use different base years (unfortunately the figures on gas demand by oblast are available only for 1991).

The figures in Table 2.7 indicate a strong degree of regional concentration in likely decline in gas demand, with more than 14 BCM available from the Central region, of which more than one-third would be from the city of Moscow. Nearly 18 BCM would be available from six oblasts in the Urals and nearly 8 BCM from four oblasts in the Po-Volga region. The grand total

---

[19] Rosgazifiatsiya data for 1991 for all oblasts in Russia. Regional production data from GOSKOMSTAT; many thanks to Professor Philip Hanson of CREES at Birmingham University for giving me access to the latter. See also Philip Hanson, *Regions, Local Power and Economic Change in Russia* (London: Royal Institute of International Affairs, PSBF paper, 1994) and 'Die wirtschaftliche Lage Ruslands realwirtschaftliche Anpassung', Deutsches Institut für Wirtschaftsforschung, *Wochenbericht* 2/95, 12 January, 1995.

Table 2.7 Likely location and origin of surplus gas demand in 2000 relative to 1991 (BCM)

Region	Industrial	Power generation	Total
Central			
Moscow	2.5	3.0	5.5
6 oblasts[a]	5.1	3.6	8.7
Regional total			14.2
Po-Volga			
Samara	2.6	0.8	3.4
Tatarstan	2.0	—	2.0
2 oblasts[b]	2.1	0.2	2.3
Regional total			7.7
Urals			
Ekaterinburg	3.2	1.7	4.9
Perm	2.6	0.3	2.9
Bashkir	2.9	1.1	4.0
Chelyabinsk	2.6	0.3	2.9
2 oblasts[c]	1.3	1.7	3.0
Regional total			17.7
Total	26.9	12.7	39.6

[a] Tula, Ryazan, Nizhnegorod, Belgorod, Kursk, Lipetsk.
[b] Saratov, Volgagrad.
[c] Orenburg, Udmurt.

of 39.6 BCM represents a little less than 50% of the total expected demand reduction shown in Table 2.6. Perhaps the most interesting figure is the potential availability of 12.7 BCM from base-load power generation, where the number of sites involved is relatively limited and all of them are relatively close to major transmission lines.

The most controversial figures in Table 2.7 are for Moscow, where the regional gas transmission company has produced a forecast which sees demand increasing steadily – in both the city and the region – from 50 BCM in 1990 to 61.3 BCM by 2000 and further to 71.6 BCM by 2010.[20] Even with significant population growth requiring major expansion in living accommodation, these projections seem much too high, at least for the period

[20] A. D. Agarkov, 'Problems of Major Cities', paper presented to RIIA Conference.

up to 2000. Mosgas, the municipal company which supplies Moscow city, is suffering from extremely high levels of non-payment, particularly from electricity generation companies. In such circumstances it is hard to see any significant increase in demand. In general terms, Moscow and St Petersburg probably have the best case to make for (electricity and) gas demand growth in the future, but this seems more likely to come after 2000.

Beyond 2000: gas demand and economic recovery
The difficulties of making even short-term projections should indicate that trying to make longer-term forecasts is a matter of pure guesswork. As noted above, the principal problem is to take a position on the future industrial structure of Russia, particularly in terms of the balance of energy-intensive industry against light industry and services. Despite analysis which suggests that Russia may have a comparative advantage in energy-intensive industries, it is difficult to imagine that these industries will regain the size and pre-eminence they enjoyed at the close of the Soviet period.[21]

Furthermore, there is general agreement that very sizeable energy conservation gains would result from the introduction and enforcement of cost-based pricing. Equally sizeable efficiency gains would result from plant replacement, using currently available technology and equipment. The total effect of even 50% of the conservation and efficiency potential being realized casts serious doubts on the likelihood of significant increases in energy and specifically gas demand in the industrial and power generation sectors during the first decade of the next century. Less certain is the demand profile of the residential/commercial/service sector of the economy, which is likely to expand enormously to achieve a share resembling that which it holds in OECD countries. However, two caveats are in order here: first, the points made above about the level of prices, and lack of payment enforcement due to political pressures and inability to disconnect customers, make it extremely difficult to project real levels of demand even at present, let alone into the future; second, the widespread use of district heating for residential customers makes comparisons with OECD countries extremely difficult.

---

[21] Dienes et al., *Energy and Economic Reform*, p. 154, citing work by Senik-Leygonie and Hughes.

The picture which emerges is of gas demand falling until around 2000. Two contrasting scenarios can then be sketched: in the first, gas demand recovers rapidly and returns to its 1990 level by 2010; in the second, gas demand recovers more slowly and remains significantly below its 1990 level even by 2010. In both cases, demand continues to fall until at least 2000. The geographical dimension of this demand reduction and the location of likely future increases is extremely important. Just as the industrial demand projections pose the challenge of trying to predict the future industrial structure of Russia, so identifying the location of future increases in gas (and electricity) demand poses the additional challenge of predicting the geographical redistribution of industrial and commercial activity. As noted above, this author's prediction would be that the Central (particularly around Moscow) and the North-West (particularly around St Petersburg) are likely to be the first Russian regions to emerge from industrial recession and that much new industry will locate (or relocate) in these regions. By contrast, the Urals region seems very likely to remain depressed for a much longer time, as new industry and commerce will require a protracted period to replace the massive heavy industrial complexes which will continue to shrink in size over the next decade.[22]

2.4 Policy issues with a significant impact on gas demand projections

We have already noted that a scenario of demand decline is not the expectation of Gazprom. Because of this, and the significant uncertainties surrounding any projections, it is important to consider other factors, and specifically policy considerations which could have a countervailing impact on gas demand. Three issues will be considered here: gas substitution in power generation; aggressive gas marketing (including a search for new markets); and continued toleration of large-scale non-payment.

Gas substitution in power generation
Clearly, future trends in demand for power in Russia will have a major bearing on future natural gas demand. This subject requires a major study in itself,

---

[22] This is an economic judgement. A more political approach might recognize the possibility that the political power of the military-industrial complex may revive in the future, and with it the economic fortunes of the Urals.

but the most important question arising from Table 2.3 would seem to be the future of nuclear and coal-fired power stations in the Central and western regions of the country. Looking at the availability of coal and the rising cost of long-distance transportation from centres of production, some contraction of coal-burning power plants might be expected in the Caucasus and also in the Central region. However, in neither case is this likely to add up to a large amount of capacity. In the Urals, which is far more dependent on coal, a switch from coal- to gas-fired plant is less likely. Indeed, given the very sharp decline in economic activity – and hence in power demand – in the Urals region, it is doubtful whether any new power generation capacity will be needed. In addition, given the relatively desperate situation of virtually the entire coal industry and the political consequences of large-scale mine closures, the government will be keen to maintain production at as high a level as possible for the foreseeable future.

Potentially more interesting in terms of gas substitution is the future of the Russian nuclear power sector. Following the Chernobyl nuclear accident, major international initiatives were mounted to improve safety at both the RBMK (Chernobyl-type) and the VVER (pressurized water) reactors.[23] There was even discussion of wholesale closure of such reactors and replacement with thermal plants (see Box 2.1 overleaf). Thus far, however, the level of both assistance and plant closure has been insignificant.

A general conclusion for the power generation sector in the years ahead would be that, while older non-gas plant (coal, nuclear and the few remaining base-load oil-fired stations) may close in the future and new gas-fired plant will be built, the pace of this process will probably not be rapid and will be determined by the fall in electricity demand. In the period up to 2000 the principal issue is likely to be that of closing stations in regions where power demand is falling rapidly or where non-payment has reached such proportions that stations cannot continue to operate. This seems much more likely than any significant programme of building new gas-fired units. The only capacity-building programme sufficiently large to make a significant and immediate

---

[23] Roberto Mussapi and Steve Thomas, *Improving Safety in Nuclear Power Generation in the Former Soviet Union and Eastern Europe: The Effectiveness of Current Initiatives* (London: Royal Institute of International Affairs, PSBF Briefing no. 2, December 1994).

difference to future gas demand would be a policy decision to sweep away all (or large parts) of the Russian nuclear power sector. In the absence of another serious incident at a nuclear station this seems highly unlikely. Even if such an incident were to occur, it is by no means certain that the investment to build new gas-fired stations would be forthcoming from either Russian or foreign sources.

New markets for gas
Even if substitution possibilities do not necessarily suggest any major increase in gas demand there remains the possibility that Gazprom may seek to expand demand either through aggressive marketing or in developing new markets. Any marketing drive might concentrate on large customers close to existing pipelines where payment can be enforced. In this connection, there have been reports of predatory pricing by Gazprom, forcing large customers to switch from coal to gas. The extent to which this is, or might become, widespread is hard to estimate. In terms of regions, penetration in the North West seems relatively low and demand here might be expected to increase rapidly. This is perhaps the only area in which significant increases in gas-fired power generation might be expected (see Tables 2.3 and 2.4).

In terms of new sectors, the prospect of large-scale use of compressed natural gas as a vehicle fuel (particularly in large trucks) has been on the agenda for a number of years. In the longer term this does seem a promising market, but it will require large amounts of capital investment to open up. The main expansion sector for gas in the short term is the residential/district heating market but, as we have seen, the problems of low prices and non-payment in this market are such that significant changes need to take place before expansion makes commercial sense.

Government policy on prices and non-payment
Possibly the most important issue in the calculation of future gas demand is how government policy on prices and treatment of consumers who refuse to pay bills will evolve. This is an economy-wide issue with major political implications, notably the potential for increasing unemployment in cities and regions dependent upon industries where enforcement of payment of bills or disconnection could result in closures.

Box 2.1 Possible replacement of Russian nuclear plant by gas-fired generating plant

The possibility exists that a further significant incident at a Russian nuclear power station might add urgency to both international and domestic initiatives for closing at least some of these stations.[a] Table 2.8 shows existing nuclear power stations in Russia. While the gross capacity of these 24 units is over 20 GW, the effective capacity is only around 19 GW. A decision to replace the entire nuclear power generation capacity with modern combined-cycle gas-fired units would give rise to an annual gas demand requirement of 30 BCM.[b] Such units could be built relatively quickly – perhaps within a five-year period – but a conservative estimate of the capital cost would be around $10 bn, even taking no account of any decommissioning expenditures for the nuclear plants.[c]

Table 2.8 Russian nuclear power plants (operational in 1994)

Site	No. of units	Type of station	Gross capacity (MW)
Novovoronezh	2	VVER-440/230	834
Kola	2	VVER-440/230	880
Kola	2	VVER-440/213	880
Novovoronezh	1	VVER-V187	1000
Kalinin	2	VVER-V338	2000
Balakova	4	VVER-V320	4000
Leningrad	4	RBMK	4000
Kursk	4	RBMK	4000
Smolensk	3	RBMK	3000
Total	24		20594

Source: Mussapi and Thomas, *Improving Safety in Power Generation* (see note [c]), tables 1–4.

An additional demand requirement of 30 BCM would be significant. Yet it is reasonable to ask:

- whether the entire Russian nuclear industry would need to be swept away in this process, given that the plants are of different designs and different ages;
- whether given falling power demand in Russia plant closures would require immediate replacement;

- whether the capital can be made available either within Russia or from abroad to build such a large amount of generation capacity within such a very short time, given that foreign investors would probably receive payment in roubles at prices paid by Russian electricity consumers.

It may be more realistic to focus on replacing certain types of reactor, for example units which were built prior to 1980. This might suggest concentrating in particular on the VVER 440/230 plants and two of the four RBMK units at the Leningrad site. Replacement of 3700 GW of capacity would suggest an additional gas requirement of around 6 BCM per year, and a capital investment of nearly $2 bn (with no allowance for decommissioning). Alternatively, attention might concentrate on replacing the RBMK (Chernobyl-type) reactors. Replacement of 11 GW would give a gas requirement of 17 BCM per year and a capital cost of $8.5 bn.[d]

There is, as yet, no sign that either Russian or foreign investors have either the financial capacity to provide or the interest in providing such large amounts of investment. To date, efforts have concentrated on safety modifications at existing reactors. The Russian nuclear industry is still thinking in terms of expansion rather than widespread plant closure. Foreign assistance in this area appears to have been modest, poorly focused and held back by issues of indemnification against potential future liability.[e] A Russian Federation–European Commission memorandum of understanding on nuclear liability signed in February 1995 may begin to breathe some life into the foreign assistance programme.[f]

Notes

[a] For early work in this area see Stewart Boyle, 'Soviet RBMK Reactors: Options for the Future', *Energy Policy*, August 1992, pp. 760–2. The present author has benefited from reading additional work commissioned by Greenpeace International on Russia, Ukraine and Slovakia.
[b] Assuming an annual gas requirement of 1.55 BCM per 1000 MW of base-load gas-fired capacity.
[c] Assuming $0.5 bn per installed 1000 MW of capacity.
[d] There are other alternatives, for example to replace the VVER 100 MW reactors. See the *Joint Energy Alternatives Study* prepared for the Energy Policy Committee of the US–Russia Joint Commission on Economic and Technical Cooperation, Washington D.C., 1994.
[e] Roberto Mussapi and Steve Thomas, *Improving Safety in Nuclear Power Generation in the Former Soviet Union and Eastern Europe: The Effectiveness of Current Initiatives* (London, Royal Institute of International Affairs, PSBF Briefing no. 2, December 1994).
[f] For details of the memorandum and the amounts of aid actually granted (as opposed to committed in principle) see 'EU/Russia Nuclear Liability Pact Puts Ball in Western Industry Court', *EC Energy Monthly*, 20 March, 1995, pp. 6–7.

   Prices and payment policy is an important area in which the gas industry
remains a state-controlled sector, notwithstanding the privatization initiatives
which have been implemented. Despite the weakness of central government,
it is clear that great care has been taken over the speed at which prices are
allowed to increase and the number of disconnections that are allowed to
occur. While Western economic commentaries tend to concentrate on raising
real prices to 'world' levels, non-payment is much the more urgent issue.[24]
For the gas sector, industrial – and specifically power generation – customers
are by far the most important problem. Prior to the March 1995 increases,
prices for the residential/district heating sector, had fallen so low that in
many areas of the country it is doubtful whether the cost of bill collection
made the exercise worthwhile. It seems certain that, given a free hand
Gazprom (and to a lesser extent the distribution companies) would have
taken a much harder line on non-payment, at least by industrial customers.
A more aggressive disconnection policy, particularly for power stations,
would in turn force the electricity sector to take a harder line with its
customers.

   The figure of 53% non-payment by Russian gas consumers in 1994 would
suggest that suppliers received no payment for around 190 BCM of gas.
Two questions arise from this which are of crucial importance for future
demand projections: first, what would have been the level of demand (at the
prices then in force) if payment had been enforced by disconnection? Second,
is large-scale non-payment sustainable in anything other than the short term?
Clearly neither question can be answered with any certainty; but, perhaps
more important, the questions point up a pertinent conceptual difference
between the Russian and the Western definitions of 'demand'. For Russians,
gas demand is what consumers *need*, for Westerners it is what consumers
can *afford*; this partly explains the continued Russian projections of high
demand figures in Table 2.5.

   The most likely outcome is that the Russian government and Gazprom
will gradually move towards the Western view of demand, the first step
being to disconnect enterprises which have no conceivable hope of paying a

---

[24] 'World' price levels are a completely meaningless term when applied to gas. In the Russian
context the term is a shorthand for (an approximation of) an average gas export price to
European countries.

fuel bill (because their output is worthless). It is important to recognize that if only 10% of the volumes delivered and not paid for in 1994 were delivered to enterprises which could be considered bankrupt with no means of paying for them, this would free up an additional 19 BCM of gas. Non-payment is probably not sustainable at the level of 1994, if only because it is likely to spread as those who have paid see no incentive for continuing to do so. The power generation sector, which is running very high levels of non-payment with gas suppliers (because if consumers are not paying their gas bill they are probably also not paying their electricity bill), is compounding the payment problem for Gazprom. There is little information as to whether non-payment is sectorally or regionally concentrated. It would be possible for the government to acquiesce in continued high levels of non-payment in certain well-defined sectors or regions of high political sensitivity; but this would prevent the implementation of any significant economic reform or restructuring.

To some extent, non-payment can be represented as a specific government policy. Indirect government subsidies are given to industries by this means, and the government forgoes taxation of the huge profits which Gazprom would be making if consumers were paying their bills in full. To the extent that the government remains in control of this process it can use Gazprom as an agent to close down enterprises by selective implementation of disconnection over a period of years. This is probably more likely than a massive and sudden payment crackdown which could lead to widespread plant closures and consequent unemployment. To the extent that non-payment is allowed to continue at anything resembling 1994 levels, any price reform will have a significantly reduced effect. Partly for this reason and partly because of the continuing subsidy given to large sections of Russian industry, it is likely that while internal oil (and possibly even coal) prices will move towards international levels, gas prices will be much slower to follow.[25] Indeed, it may be that policy towards non-payment will provide a useful yardstick against which to measure successive Russian governments' continuing commitment to the reform and restructuring of industry.

---

[25] The other major difference is that, unlike oil (and to a lesser extent coal) exports, gas exports are constrained both by transportation capacity and by inelastic demand in export markets. We return to this issue in Chapter 3.

From this morass of uncertainty two conclusions emerge for projections of Russian gas demand: first, disconnecting consumers who do not pay their bills could reduce demand by a minimum of 20 BCM per year; second, to the extent that action is not taken to disconnect consumers who do not pay their bills, there is no economic rationale for investing in new sources of supply.

For a variety of reasons, this chapter has foreseen a significant fall in gas demand over (at least) the next five years which may have major potential consequences for Russian gas export availability. But given the nature of gas transportation, no easy generalizations can be made about translating internal availability of gas into exports.

Chapter 3

The Export Dimension: Sales, Transit and the Expansion of Pipeline Capacity

3.1 The legacy of the Soviet era

By the end of the Soviet era a gas export trade of more than 100 BCM/year had developed between the USSR and European countries. As we noted in Chapter 1, the development of exports was closely related to the development of individual fields – Orenburg, Urengoy, Yamburg – with dedicated pipelines from those fields to export markets. During the Soviet era two distinct markets for gas exports evolved: the Central/East European countries which were members of the Council for Mutual Economic Assistance (CMEA), and the markets of OECD Europe. These markets were characterized by different contractual relationships and different currencies of payment. While the details of relationships with individual countries are complex, the essence of the contract was that Central and East European countries received gas under long-term government-to-government 'compensation agreements' as repayment for their involvement in the construction of gas facilities within the USSR. They also purchased gas under agreements where prices and volumes were renegotiated annually. All gas trade was conducted on a transferable rouble or barter basis. Countries in OECD Europe, on the other hand, purchased gas on long-term (25-year) contracts which included high levels of 'take-or-pay' commitment. Many such contracts, particularly in the early years, included a significant element of counter trade, with government-sponsored credit packages supporting exports of large-diameter steel pipe and (turbines for) compressor stations by companies located in the countries which were importing gas. By the close of the Soviet era, hard-currency earnings from gas exports had grown to around 20% of Soviet merchandise exports.[1]

---

[1] For the history of natural gas exports in the Soviet era see Jonathan P. Stern, *Soviet Natural Gas Development to 1990* (Lexington: Lexington Books, 1980); Javier Estrada et al., *Natural Gas in Europe: Markets, Organisation and Politics* (London: Pinter, 1988), pp. 169–88.

The break-up of the Soviet Union created a set of trading relationships between newly emerging sovereign states which had formerly treated these relationships as internal transfers without financial significance. The problems of gas trade between former Soviet republics, and particularly the triangular relationship between Russia, Turkmenistan and Ukraine, not only affect Russia's relations with the other republics, but are crucial to the trading relationship with European countries.

3.2 Trade with former Soviet republics

The statistics in Table 3.1 are not entirely reliable and contain a certain amount of interpretation on the part of the author. Nevertheless, they can be regarded as a reasonably accurate picture of Russian gas deliveries both before and immediately following the break-up of the Soviet Union. What these figures do not show is the pattern and extent of inter-republic trade during the Soviet period. The essentials of this trade were that Russia supplied Ukraine, Belarus, Moldova and Kazakhstan, while receiving gas from Turkmenistan and Kazakhstan. Turkmenistan supplied Kazakhstan and Uzbekistan and (via Russia) Ukraine, Georgia, Armenia and Azerbaijan, as well as Russia itself. In Central Asia, Turkmenistan was (and remains) the major exporter, but Uzbekistan supplies Kazakhstan, Kyrgyzstan and Tajikistan. Kazakhstan exports small quantities of gas to Russia. The exact details of exchanges, particularly between the Central Asian republics and between the latter and Russia, are especially complex.

For our present purposes, the key issue in terms of *supply* is, and continues to be, that outside Central Asia, republics receive virtually their entire gas supply either from Russia, or from Turkmenistan through the Russian system. The exception is Ukraine, which receives gas from both republics to add to its substantial (but dwindling) domestic production. The key issues in terms of *transit* are that all Turkmenistan's gas exports outside Central Asia pass through Russia which puts the latter in complete control of around three-quarters of Turkmenistan's exports; and all Russian gas exports to Europe pass through former Soviet republics, principally Ukraine but also Moldova and Belarus.

The break-up of the USSR caused immediate and continuing problems in gas trade between all of the republics, primarily taking the form of demands

Table 3.1 Russian natural gas exports to former Soviet republics, 1990–1994 (BCM)

	1990	1991	1992	1993	1994[a]
Ukraine	60.8	60.7	77.3	54.9	57.1
Belarus	14.1	14.3	17.6	16.4	14.3
Moldova	2.3	2.5	3.4	3.2	3.0
Georgia					0.4
Lithuania	6.1	6.0	3.2	1.9	2.0
Latvia	3.4	3.2	1.6	1.0	1.1
Estonia	1.7	1.9	0.9	0.4	0.8
Kazakhstan			1.7	1.1	0.6
Total[b]	92.0	90.0	106.4	78.6	79.1

[a] Preliminary figures.
[b] Totals do not add owing to rounding.
Sources: David Cameron Wilson, *CIS and East European Energy Databook 1994* (Tadcaster: Eastern Bloc Research, 1994); *European Gas Markets*, various issues.

for gas and transit tariffs to be paid for at prices, and in currencies, which (with the exception of Russia) none of the recipients can afford. Refusal and/or inability to pay has resulted in the amassing of huge debts to suppliers (principally Russia and Turkmenistan) and periodic cutbacks in supply because of non-payment. These cutbacks have caused not only hardship but, in cases where gas has been in transit to third countries, the diversion of supply intended for those countries. While we are principally concerned in this chapter with the Ukrainian situation, it is important to recognise that similar problems have arisen in Central Asia and the Caucasus.[2]

In terms of relationships with Russia, the former republics can be divided into three groups: the Baltic states; Ukraine, Belarus and Moldova; and the Central Asian countries.

The Baltic states
Latvia, Lithuania and Estonia were first to leave the Union and as a consequence were, by 1992, required to pay prices which (in terms of level and currency) resembled European border prices. These higher price levels,

[2] For example in the trade between Turkmenistan, Kazakhstan and Uzbekistan; and also in exports from Turkmenistan to Georgia, Armenia and Azerbaijan.

combined with economic restructuring and the availability of lower-cost (principally Russian) gas oil and high sulphur fuel oil, caused Russian gas deliveries to these countries to fall from 11.1 BCM in 1991 to 3.3 BCM in 1993. In 1994, some recovery of volumes occurred (see Table 3.1). With some flexibility on the part of Gazprom and the Russian government in the area of pricing it is likely that deliveries to all three countries will continue to recover in the future. Russia's desire to use the storage facilities in Latvia will be a useful bargaining counter in gas trade relationships. Nevertheless, the reason why this study includes these countries under 'former Soviet republics' rather than 'exports to Europe' is that, in terms of tolerance of debts, they are clearly still regarded by Gazprom as having yet to make the transition to full European status.

Ukraine, Belarus and Moldova

The story of Russia's relationships with these countries since the break-up of the Union is one of mounting gas debt in the face of bills which these countries have no money to pay.

Ukraine   In volume terms, Canadian gas exports to the United States constitute the only other bilateral gas trade worldwide which comes close to the volume of Russian exports to Ukraine. It is only possible to understand the magnitude of this trade by recognizing that during the 1990s Russian deliveries to Ukraine have been the equivalent of 55–80% of its entire export volume to Europe. Another way of looking at these volumes is that during this period they have fluctuated between the annual gas consumption of Italy and that of Germany.

Russia's position *vis-à-vis* Ukraine is extremely vulnerable in that more than 90% of its gas exports to Europe pass through that country. On more than one occasion the Ukrainian gas industry has taken gas from the export pipelines for its own needs (see Box 3.1). Gazprom's freedom of action is severely constrained. From a commercial standpoint, it cannot be seen to allow Ukraine to disregard the financial aspects of the relationship completely for two reasons: first, the volumes involved are so large that they simply cannot be ignored; second, it is essential that progress be made towards some longer-term contractual commercial framework that Ukraine is willing and able to honour. Such progress is unlikely while Ukraine believes that it can get away with paying only a part of the agreed price for Russian gas. The other side of Gazprom's commercial calculation is that any disruption in European exports – for however short a time, and even if it has no noticeable impact on European gas consumers – does damage to Russia's reputation as a secure supplier of gas to its convertible currency markets.

This is why, despite the debts and periodic diversion of gas supplies, Gazprom has been extremely conciliatory towards Ukraine over the past three years. The Russian position has contrasted sharply with that of Turkmenistan, which has cut deliveries for long periods due to non-payment and disputes over prices. This has periodically left Russia in the position of having to give extra gas to Ukraine to make up for non-deliveries from Turkmenistan, as well as delivering under its own bilateral arrangements. The clearest instance of this occurred during the seven-month cessation of supplies in 1992, due to a price dispute between Kiev and Ashkhabad. Table 3.1 shows a 17 BCM increase in Russian supplies to Ukraine to compensate; smaller compensatory deliveries have occurred in 1993 and 1994.

Understandable Ukrainian political sensitivity towards Russian influence has been a considerable obstacle to a partial commercial solution which may lie in Gazprom taking some degree of ownership in Ukrainian gas transmission and storage assets. A large part of Ukraine's 35 BCM of storage capacity was built for the specific purpose of ensuring security of Soviet, now Russian, exports to Europe. While the Ukrainian government seems prepared to countenance Russian equity shares in Ukrainian enterprises which are primary customers for gas (such as metallurgical and chemical plants) it is not prepared to allow Russia to acquire a controlling share, and is trying to limit the proportion of existing gas debt which can be exchanged for equity shares.[3] The Ukrainians are particularly sensitive as regards equity interests in actual gas assets (pipelines and storage facilities), which are Gazprom's primary targets, indicating that shares may be limited to 15–20%.

Gas deliveries are now an important issue in the political and security relationship between Russia and Ukraine, having featured in the package of

---

[3] 'Ukrainian President Backs Equity-for-Debt Deal for Russian Gas', *BBC Summary of World Broadcasts* (henceforth *SWB*), Part 1, SUW/O376 WD/4, 24 March, 1995; Chrystia Freeland and Matthew Karwiski, 'Natural gas fuels Kiev's row with Moscow', *Financial Times*, 8–9 April, 1995.

Box 3.1 Ukrainian diversions of Russian gas exports in transit to European countries

Since the break-up of the Soviet Union much publicity has been given to interruptions of Russian gas supplies to European countries due to unauthorized diversions of the volumes which pass through Ukraine. Diversion of supplies by the Ukrainian gas industry has been the product of a complex set of circumstances and negotiating positions in which the principal factors are accumulated debt to Russian and Turkmen suppliers and the degree to which non-payment merits cutbacks in these supplies.

The most serious disruption in Russian supplies to Europe occurred in October 1992, when for a period of around 10 days supplies to Germany were 20–50% below contracted levels.[a] The fact that Turkmenian supplies to Ukraine had been stopped some months previously because of a price dispute and that debts to Gazprom had risen to very high levels may explain (but does not excuse) Ukrainian actions.

In March 1993 there were (unconfirmed) reports of Ukrainian diversion of supplies, but the next confirmed events were in September 1993 when gas destined for Turkey, Romania and Bulgaria was reduced substantially.[b] In February 1994 the debt crisis between Russia and Ukraine again caused Russia to suspend a portion of deliveries to some Ukrainian customers and this again caused unauthorized diversions which amounted to around 20% of European supplies; importers in France, Germany and Italy registered the fall in deliveries.[c]

In order to try to stabilize the situation and avoid any similar problems in the future, Gazprom signed an agreement in April 1994 whereby it would deliver 10 BCM of gas to be stored in two Ukrainian facilities for delivery to European customers in the winter months.[d] In addition, one of the major Ukrainian transmission companies stated that it would not interfere with gas in transit to Europe.[e]

In November 1994 it was reported that deliveries to Europe were once again reduced.[f] However, the Ukrainians denied that they had diverted the gas and explained the problem in terms of shortfalls in supplies from Turkmenistan due to a pipeline accident in Central Asia.[g] But by the first week of December, Ukraine acknowledged that it had responded to Russian cutbacks (because of failure to meet debt repayments) by diverting around 20% of European supplies in order to keep Ukrainian industries running.[h]

From these rather sparse details a few general points can be made: first, so far these episodes have not involved a complete *interruption* of deliveries to Europe, but rather a *reduction* of deliveries which (in one case) reached 50% of one importer's supplies; second, these reductions can be measured in terms of days and only one episode appears to have exceeded a week; third, there has always been ample warning of these reductions, allowing importers to make other arrangements.

The diversions tend to take place in spring and winter, at times when Ukraine needs to replenish its storages. No diversions have taken place during a period of severe weather in Europe when importing gas companies might be seriously stretched in terms of supplies. These two factors lead to the possible conclusion that the Ukrainians are prepared to use diversion as part of their supply management process – and part of their negotiating process with the Russians – but that they are well aware of the impact which serious disruption might have in Europe and would not knowingly risk such an event.

Up to the present, the problems have been sufficiently serious to make importers nervous and to have caused minor inconvenience on a small number of occasions. While this situation is not satisfactory – and shows no sign of being swiftly resolved – it cannot be said to constitute a major security threat to European gas supplies.

Notes

[a] 'Does Russian Gas Curtailment Signal a Decline in Reliability?' *Gas Matters*, October 1992, p. 14.

[b] 'Russian Suspends Gas Supplies to Industry in Eastern Ukraine', *BBC Summary of World Broadcasts* (henceforth *SWB*), Part 1, SUW/0301 WD/2, 1 October, 1993.

[c] 'Russian Cuts to Europe Cloud Euro Gas Market', *International Gas Report*, 4 March, 1994, pp. 1–2.

[d] 'Ukraine and Russia Sign Accords on Storage of Russian Gas in Ukrainian Depots' and 'Russia's Gazprom Steps Up Gas Deliveries to Ukraine for Export to West', *SWB*, SUW/0329 WD/2, 22 April, 1994 and SUW/0333 WD/1, 20 May, 1994.

[e] 'Ukraine Won't Cut Off Gas to Western Europe, Official Says', *SWB*, SUW/0349 WD/2, 9 September, 1994.

[f] One report suggested that supplies fell to one-quarter of intended levels: 'Russia Agrees to Restore Gas Supplies to Ukraine', *SWB*, SUW/0362 WD/8, 9 December, 1994.

[g] 'Ukraine Denies New Gas Grab', *International Gas Report*, 25 November, 1994, p. 11.

[h] 'Russia/Ukraine Row Again Hits Other Buyers', *International Gas Report*, 9 December, 1994, p. 3.

agreements which have included issues such as the future of the Black Sea Fleet and Ukrainian nuclear weapons.⁴ The inter-governmental agreement signed between the countries in February 1994 which sets out the terms of sale and transit for a ten-year period may be a useful reference-point in this relationship. Nevertheless a sound commercial basis for this huge trade, with its clear mutual dependence aspects, has yet to be established.

Belarus After Ukraine, all other trades pale into insignificance in volume terms; but, at 14–17 BCM per year, Russian deliveries of gas to Belarus are by no means negligible; in addition, Belarus provides transit for up to 7 BCM per year to Poland. But the importance of Belarus as a transit route for Russian exports will increase significantly in the future. Largely because of the need to diversify away from dependence on Ukraine (but also for other reasons), a second export corridor is being established through Belarus and Poland to Germany.⁵ This pipeline route will be known as the 'Yamal pipeline', despite the fact that, as noted in Chapter 1, the line will be supplied with gas from existing fields through existing pipelines as far as the Russian border. For the first phase of deliveries (scheduled to start in late 1996) only the new Belarussian and Polish sections of the line need to be built. Russian sources speak of building the pipeline 'from the market to the fields'. That is to say, they will only build additional pipeline and production capacity as it is required to develop the two lines with an eventual capacity of 65.7 BCM.⁶

The Russian relationship with Belarus is relatively stable and is underpinned by a January 1994 inter-governmental agreement which sets out the sale and transit relationships between the countries over a 20-year period. Under this agreement – which in early 1995 had still not been ratified by the Belarus parliament – the Belarussian transmission company Beltransgaz would

---

⁴ 'Russia–Ukraine: The Peace Dividend Is Gas', *Gas Matters*, January 1994, pp. i–iii.
⁵ The other reasons for creating a new corridor relate to the physical security of having such a large number of pipelines converging on such a relatively small area. In the event of a major explosion, the entire capacity through Ukraine could be affected. Also, in comparison with the Ukrainian route, the Belarus–Poland route is significantly shorter in distance to North European countries.
⁶ The protocol between Russian and Poland foresees volumes starting to flow in 1996, increasing to: 32.3 BCM by 1999, 62.7 BCM by 2005 and 65.7 BCM by 2010. 'Gazprom swaps Yamal dreams for a more mundane reality', *World Gas Intelligence*, 28 April, 1995.

become a subsidiary of Russia's Gazprom. This apparently means that the assets of Beltransgaz have been transferred to Gazprom under a 99-year lease.[7] In return, Russia is said to have agreed to double its deliveries to Belarus by 2010, although it is not clear where the country would use such large quantities of gas. A more plausible outcome, at least in the short term, is that Belarus would use gas received from transit payments from the Yamal line to reduce its existing gas bill from Russia.

Moldova   With an annual gas requirement of only 3–4 BCM, it is easy to forget Russian sales to Moldova. However, the country is also an important transit route along which all gas to south-eastern Europe (Romania, Bulgaria and Turkey) passes. At the end of 1994, it was reported that Gazprom had set up a company with Moldovagaz which would jointly own the country's gas assets along the same lines as its arrangement in Belarus.[8]

At 1 January 1995 the gas debts of former Soviet republics were estimated by the Russian government at 7.76 trillion roubles (approximately $1.9 bn), from Ukraine (more than 5.48 trillion roubles), Belarus (around 1.51 trillion), Moldova (625 billion), Latvia (98 billion), Estonia (30 billion) and Georgia (14 billion).[9] What we have tried to show above is that Russian management of these debts will need to be dictated by the transit aspect of the relationships with the former Soviet republics. Transit is so important for Russia that it overshadows its entire commercial and political relationship with these countries.

The Central Asian countries
Turkmenistan   As noted above, during the Soviet era Russia received significant quantities of gas from Turkmenistan. Immediately following the break-up of the Union, Turkmen demands for payment from Russia in hard currency at 'world prices' resulted in sharp reduction of Russian purchases. In addition, within the framework of the Soviet Union, when the republic

---

[7] 'Russia/Ukraine Fix New Gas Debt Deal', *International Gas Report*, 15 April, 1994, p. 2.
[8] 'Moldova's Gas Imports from Russia Will Cost Less in 1995', *SWB*, SUW/0365 WD7, 6 January, 1995.
[9] 'Russia Says Former Soviet Republics Owe it R9000 bn for Fuel', *SWB*, SUW/0370 WD/5, 10 February, 1995.

produced around 10% of total Union production, Turkmenistan was allowed a similar share in the hard currency revenues earned from sales to Europe. With the break-up of the Union, Russia's Gazprom became the legal entity holding the European export contracts. Turkmenistan's hard currency quota was reduced in 1992 and then discontinued in 1994 (see Table 3.2 below) leaving the country in the position of trying to find alternative export routes to Europe via Iran and Turkey, as well as exploring alternative export markets in Pakistan, China and Japan.[10] The Turkmen position is perhaps more precarious than the country's behaviour would suggest, given that it is totally dependent on Russian transit for the majority of its export markets. Unless the Russian–Turkmenian relationship improves in the future, small changes in inter-republic trade – such as the beginning of Russian supplies to Georgia – should be of concern to Turkmenistan (see Table 3.1).

In late 1994 there was significant activity suggesting that the Russian–Turkmen gas relationship was being moved on to a different basis, or possibly that some resumption of the previous relationship was imminent. Press reports gave transit volumes agreed for 1995 between the Turkmen government and Gazprom for Ukraine, Azerbaijan, Armenia and Georgia. Most significantly, the agreement is said to include a sale of 10 BCM of Turkmen gas to Russia in 1995.[11] With the current state of intra-CIS politics and ability to pay, nobody can say whether these arrangements will have any practical significance. However they do suggest that some degree of order is being restored in gas commerce between Turkmenistan and its customers, and that Russia may be willing – despite its own oversupply of gas – to resume Turkmen purchases, even if only to deliver the gas to Ukraine on its own account. From the Turkmen perspective, a desire to re-establish orderly trading relationships may stem from the significant decline in production because of a curtailment of export markets – and hence foreign exchange revenues – since the break-up of the Soviet Union.[12] It is not clear whether Turkmenistan understands that ambitious

---

[10] For an excellent summary of the projects which have been advanced see James P. Dorian, Ian Sheffield Rosi, S. Tony Indriyanto, 'Central Asia's Oil and Gas Pipeline Network: Current and Future Flows', *Post-Soviet Geography*, vol. 35, no. 7, 1994, pp. 412–30.

[11] It is not known whether there is any hard currency component to this sale. See 'Turkmenistan and Russia Finalise Gas Transit Deal', *SWB*, SUW/0366 WD5, 13 January, 1995.

[12] In 1991 Turkmenistan produced nearly 80 BCM of gas and exported over 70 BCM to other Soviet republics; in 1994 its production was only around 35 BCM.

plans to pipe gas through Iran and Turkey to Europe will not be realized in anything other than the very long term.

Kazakhstan Russian trade with Kazakhstan has not been particularly significant in either gross or net terms. By 1994 this trade was reported to have been reduced to negligible proportions, principally because of payment problems. This may change in the future with the inclusion of Gazprom in a joint venture to develop the Karachaganak gas field (with around 1 TCM of reserves) in northern Kazakhstan. Thus far the agreement, with partners British Gas and AGIP (the production subsidiary of the Italian energy conglomerate ENI), relates only to restoring current production capacity at the field (which is relatively small), but if successful this could pave the way for a major export project through Russian pipelines (potentially) to Europe.[13]

The general interaction of Russian and Central Asian gas trade in the future will be extremely important, given the vast resources in both Turkmenistan and Kazakhstan and the existence of pipelines through Russia to bring significant volumes to European markets either directly or by displacement. We return to this question in Chapter 4.

In summary, the future of Russian gas relationships with the former republics is not straightforward and cannot be easily generalized. The problems which have been created by the break-up of the Union in the established interdependence relationships require sensitive handling both by Russian foreign policy-makers and by Gazprom. Attempts by Russia to reimpose Soviet-type 'imperial' power will probably lead to serious physical hardships in the former republics, and serious financial suffering on the Russian side. Cooperation could lead to a 'win–win' solution with both sides benefiting from stable transit and sales arrangements. These arrangements are more likely to be understood and welcomed by gas industry executives than by politicians with different and wider agendas.

In terms of Russian export volumes to the former republics, the signs are mixed. From the point of view of indebtedness it would be natural to draw

---

[13] For details of the agreement see 'Karachaganak Edges Cautiously Forward', *Gas Matters*, March 1995, p.1. For the text of the agreement see *Concerning the Agreement between the Governments of the Russian Federation and the Republic of Kazakhstan on Cooperation in development of the Karachaganak Oil/Gas/Condensate Field* (Moscow: Russian Federation Government, Resolution no. 173, 23 February, 1995).

the conclusion that deliveries will do no better than remain stable, and will probably fall. But this is to ignore the interdependence relationships noted above. Not only may the Russian government consider it has a social and political obligation to maintain deliveries to the 'near abroad' at levels that avoid serious hardship, but in the current market environment cutting back deliveries to these countries simply increases the size of Gazprom's gas 'bubble'.

In all likelihood exports will fall in the short term, but probably because of falling demand due to economic conditions in the former republics similar to those noted for Russia in Chapter 2. Until economic conditions allow the former republics to pay prices which correspond to some kind of commercial reality, Gazprom will encourage its customers to reduce their deliveries to volumes which can be considered largely as a payment for transit rights and joint ownership of transportation assets. However, joint ownership of assets carries with it dangerous political overtones and may be resisted. In any event, Gazprom has no option but to be conciliatory and attempt to move future trade to a profitable, debt-free commercial basis within a five-year period.

3.3 Exports to Europe

Table 3.2 shows Soviet/CIS gas exports to Europe since the high point of 1990, when deliveries reached 109 BCM. By 1994 exports had still not regained the 1990 level, principally due to economic recession in Central and Eastern Europe and as a response to the higher prices and payment in convertible currencies required by the Russians. Nevertheless, by the end of 1994 there were indications that exports to Central and Eastern Europe were beginning to increase again. Exports to OECD Europe stagnated, certainly during the early years, with a reorientation of OECD buyers towards alternative suppliers, perhaps due in part to non-deliveries because of Ukrainian interruption of supply.[14]

---

[14] For example, in comparison to an increase in Soviet/CIS exports to continental OECD Europe of just over 3% in the period 1990–93, Dutch gas exports increased by 25%, Norwegian exports by 15% and Algerian exports by nearly 10%. Rueil Malmaison, Cedigaz, *Natural Gas in the World*, 1992 survey tables 21 and 22, 1994 survey table 36.

Table 3.2 CIS gas exports to Europe, 1990–1994 (BCM)

	1990	1991	1992	1993	1994[a]
Source of exports					
Russia	96.0	89.6	87.9	91.3	105.8
Turkmenistan[b]	13.0	15.6	11.2	9.6	—
Destination of exports					
Former Yugoslavia	4.5	4.5	3.0	2.7	2.2
Romania	7.3	5.4	4.4	4.6	4.5
Bulgaria	6.8	5.7	5.3	4.8	4.7
Hungary	6.4	5.9	4.8	4.8	5.3
Poland	8.4	7.1	6.7	5.8	6.0
Czech/Slovak republics	12.6	13.7	12.8	13.2	14.0
Total Central/Eastern Europe	46.0	42.3	37.0	35.9	36.7
Turkey	3.3	4.1	4.5	5.0	5.0
Finland	2.7	2.9	3.0	3.1	3.4
Austria	5.1	5.2	5.1	5.3	4.7
Switzerland	0.3	0.4	0.4	0.4	0.6
France	10.6	11.4	12.1	11.6	12.2
Italy	14.3	14.5	14.1	13.8	13.7
Germany	26.6	24.4	22.9	25.8	29.5
Total OECD	63.0	62.9	62.1	65.0	69.1
Grand total	109.0	105.2	99.1	100.9	105.8

[a] Preliminary figures.
[b] Turkmenistan 'quota' for which hard currency is received.
Source: Gazexport.

Since 1990 Gazprom has developed a new strategy towards selling gas in Europe. The traditional method of sales in the Soviet era had been to sell at the border of the importing country, with Gazprom having no involvement in the transportation or marketing of the gas. However, the Soviets came to believe that by selling at the border, they were being deprived of a share in significant profits which were being made in the downstream market. Ambitions to rectify this situation crystallized in October 1990 with a joint venture between Gazprom and the German company Wintershall which has subsequently built major pipelines (MIDAL and STEGAL) in Germany and established a marketing company in Germany (Wingas), and a trading company operating elsewhere in Europe (WIEH).[15]

Gazprom has clearly stated its intention to apply this German model in every country where it sells gas. Table 3.3 provides a list of the joint venture marketing companies (plus related ventures) which have been set up so far. It will be of interest to see whether these new institutional arrangements for selling gas will in fact bring greater profits to Gazprom. The critical point for our purposes is that Gazprom clearly believes this to be the likely outcome. What these arrangements will certainly achieve is to give Gazprom first-hand knowledge of the commercial details in all the markets in which it is operating.

The Russians have shown considerable flexibility in their joint venture arrangements, sometimes taking as much as a 50% share, sometimes as little as 25%. The general principle seems to be that the new organization should be responsible for any additional volumes which are sold over and above long-term contracts. Gazprom has been scrupulous about honouring these contracts. In some cases, when the latter expire they may be taken over by the new trading house/joint stock company; in other cases the traditional buyer remains the custodian of the gas being sold under the contract extension.

In important respects, Gazexport (the division of Gazprom which handles European exports) is at a turning-point in its contractual strategy. In Central and Eastern Europe virtually all the existing contracts will expire by 1998, by which time the entire basis of the trade will have to be renegotiated. In OECD Europe Gazexport faces the difficult task of selling gas from the Belarus–Poland pipeline relatively quickly, and possibly on a different contractual basis from that which has existed up to the present. The clear interest in adding value to gas exports by moving downstream could in the future be reflected in independent power projects and a range of other ventures which would bring profits from the sales of the resulting products. Although there has been considerable discussion of such projects, few concrete ventures

---

[15] The story of the battle for German markets between Wintershall and the forces of Ruhrgas in the western part of Germany and the Ruhrgas-led VNG consortium in the eastern part will merit a separate chapter in the history of the Germany gas industry. For the origins of this battle see Jonathan P. Stern, *Third Party Access in European Gas Industries: Regulation-Driven or Market-Led?* (London, Royal Institute of International Affairs, 1992), pp. 88–91.

Table 3.3 Gazprom's joint venture marketing companies

Partner country[a]	Joint venture partner	Joint venture name	Gazprom share %
Austria	OMV	GVH (Gaz und Varenhandelshaus mbH)	
Finland	Neste	Gasum	25
France	France	Fragaz	
Germany	Wintershall[b]	WINGAS	35
		WIEH[c]	50
Greece	DEPA	Prometheus Gas	
Hungary	MOL/Mineralimpex		
	DKG East	Panrusgas	44
Italy	SNAM	Promgaz	
Poland	POGC	Gaztrading	
	POGC	Europol Gaz[d]	48
Romania	Romgaz/WIEH	Wirom	
Slovenia	Petrol	Tagdem	

[a] Trading houses have also been set up in Bulgaria and Turkey.
[b] Wintershall is a wholly owned subsidiary of BASF.
[c] WIEH is a trading company selling Russian gas outside Germany (principally in Romania and Bulgaria).
[d] Europol Gaz is the joint venture company which will build the Polish section of the Belarus–Poland ('Yamal') pipeline.

have progressed. In view of the interest, particularly in gas-fired power generation opportunities in Europe, this is surprising.

In terms of foreign investment, Gazprom's clear interest is again in the downstream market. It has repeatedly denied any interest in involving foreign investors in upstream projects, except in so far as it accepts that Russian technology and expertise are not equal to the task. This has tended to mean that the only upstream joint ventures which have progressed are those involving recovery of liquids from gas fields with, perhaps, a small quantity of gas production attached to the liquid recovery.[16] While Gazprom is

[16] For example, on the joint venture with Bechtel for liquids production at Novo-Urengoy see Ross J. Connelly, 'Northgas Ltd: Joint Venture between Gazprom and Bechtel Energy Resources Corporation', paper presented to the conference on 'Natural Gas: Trade and Investment Opportunities in Russia and the CIS', Royal Institute of International Affairs, London, 13–14 October, 1994.

interested in pure gas projects – such as that at Shtokmanovskoye in the Barents Sea – which require foreign technology, it is clearly not in a hurry to see these projects move ahead.

3.4 Transmission capacity

Any expansion of exports will depend on the creation of transmission capacity to European markets. The current situation in respect of availability of existing pipeline export capacity is extremely complicated, and none of the published material seen by this author can be considered an accurate statement of the present position. Table 3.4 shows two estimates of 'nameplate' (i.e. theoretical maximum) pipeline capacity: deliverability to the border of the CIS (Ukraine, Russia and Belarus), and capacity of existing export systems. In the former case, the total figure of 126.5 BCM is tempered by a note in the source which states that according to Russian officials 'present (1993) capacity does not exceed 115 BCM'.[17]

The December 1993 decree on gas deliveries by Gazprom used the figure of 116.8 BCM for a 1994 export quota to European countries.[18] Given that exports would always have been considerably less than this volume, it is a reasonable guess that 116.8 BCM represented the maximum technical capacity of the export system to Europe in 1994.

To delve further into the detail of the system becomes problematic. The exit capacity figures in Table 3.4 have undergone some creative manipulation by the author to arrive at a total figure of 116.8 BCM. This manipulation has been informed by the actual 1990 export deliveries of 109 BCM. It is not credible that these volumes were the maximum which could have been delivered; deliveries to Finland were significantly below maximum capacity in that year and Turkish volumes were still building to plateau level. Even these figures do not appear to give sufficient capacity to the Northern Lights pipeline, which in 1990 delivered 14.1 BCM to Belarus (see Table 3.1), and a total of 11.1 BCM to Poland and Finland.[19]

---

[17] Marie Françoise Chabrelie, *European Natural Gas Trade by Pipelines* (Rueil Malmaison, Cedigaz, July 1993), table 10, p. 6.

[18] 'On Ensuring Reliable Gas Deliveries for Consumers by the GAZPROM Joint Stock Company in 1994–96'; for translation see *Gas Matters*, February 1994, pp. 13–18.

[19] It is possible that Belarus received some gas from another pipeline from the south.

Table 3.4 Approximate 'nameplate' capacity of CIS gas export lines to Europe (BCM)

(1) Pipelines commencing in Russia and CIS

Name of line	Point of export	Route	Capacity to CIS border
Bratstvo	Uzhgorod	Czech/Slovak Republic, Hungary	3.5
Shebelinka/Izmail	Izmail	Romania, Bulgaria, Turkey	20
Northern Lights	Uzhgorod	Finland, Poland	22
Soyuz (Orenburg)	Uzhgorod	Eastern/Western Europe	27
Urengoy	Uzhgorod	Western Europe	27
Progress (Yamburg)	Uzhgorod	Eastern Europe	27
Total			126.5

(2) Pipelines outside CIS

Name of Line	Point of export	Route	Exit capacity
Northern Lights	Imatra	Finland	4.0
Northern Lights	Brest	Poland	3.8
	Kobrin	Poland	7.0
Shebelinka/Ismail	Ismail	Romania, Bulgaria, Turkey	20.0
		Hungary/Serbia	4.0
Transgas (4 lines)	Velke Kapusany	Slovak/Czech Republic	75.0
Total			116.8

Sources: Marie Françoise Chabrelie, *European Natural Gas Trade by Pipelines*, Rueil Malmaison, Cedigaz, July 1993, table 10, p. 6; PH Energy Analysis, *Petroleum Economist*, Special Supplement, September 1993, p. 40.

Despite this inability to be sure of the detail of existing capacity, the broad outlines of the future seem clear. There is a distinct intention to increase Russian export capacity through Ukraine by around 30 BCM over the next decade. Over roughly the same period the Belarus–Poland corridor will be established with a capacity of around 65.7 BCM.[20] Therefore, the intention is that by 2005–10 Russian export capacity to Europe will have been raised to more than 200 BCM per year.

---

[20] See note 6 above. The protocol Russia and Poland anticipates that export volumes to Europe (excluding the Polish imports from the line) will increase from 0.4 BCM in 1996 to 29.3 BCM in 1999, 38.4 BCM in 2000, 50.2 in 2003 before finally reaching maximum throughput of 51.7 BCM in 2004.

Looking to the future, there is likely to be a considerable incentive to export additional gas to fill the capacity of pipelines which are due to be built. This would suggest roughly a doubling of Russian exports over the next 15 years. The traditional question is which gas fields will need to be opened up to meet such a large increase in exports. Given the considerable additional availability of gas which has been foreseen in Chapter 2 – at least until 2000 – the more interesting question is where the markets for such a large quantity of gas could open up in Europe over this relatively short period of time.

Chapter 4

The Gas 'Bubble' and Beyond: Scenarios to 2010

Chapters 1–3 of this study have assembled evidence to suggest that the next five years will see a unique combination of circumstances arise from developments which have not been deliberately created, do not appear to have been foreseen and will not necessarily be welcomed by the Russian gas industry. However, these circumstances have the potential fundamentally to change commercial calculations in favour of rapidly increasing the availability of Russian gas exports to Europe at costs not significantly greater than those of current supplies. The availability of these exports could have a significant impact on European gas markets.

4.1 The period up to 2000: the emerging bubble

Two principal developments have occurred, independently of each other but both as a consequence of the break-up of the former Soviet Union. The first is the collapse of Russian industrial production in the wake of economic reform and restructuring. This collapse, which began in 1990, had by the middle of the decade begun to feed through into Russian internal energy and natural gas demand. In 1994 the size of the Russian 'gas bubble' – defined as production which remains shut in because of the lack of markets – was around 30–40 BCM. The second is the threat to the security of Russian exports of gas to Europe that has emerged because of transit problems through Ukraine. The inability of Ukrainian customers either to pay agreed prices for Russian gas, because of economic difficulties, or to forgo deliveries, because of dependence on Russian gas, has led to sporadic – but thus far not serious – reductions of deliveries to European customers (see Box 3.1 above). Nevertheless, the resulting damage to the credibility of Russia as a secure supplier to Europe has added urgency to the creation of an alternative export 'corridor' through Belarus and Poland. While the first pipeline in this new corridor has been christened the 'Yamal' pipeline, there is no necessary link between the name of the line and the origin of the gas which will be exported through it.

Under 'normal circumstances' the building of a major new export pipeline would have been preceded by up to five years of contractual negotiations resulting in the entire capacity of the line being sold on 25-year contracts, requiring high levels of 'take-or-pay' commitment from the buyers. After the signing of contracts there would be a relatively long lead time for the line to be built, plus a period during which volumes would build up to plateau levels. For the export phase of the Urengoy pipeline the period from the start of negotiations to reaching the plateau of delivery volumes was around 10 years.

For the Belarus–Poland pipeline, the circumstances are different. Because of the problems with Ukraine, the Russians are in a hurry to create additional capacity, while traditional buyers are not in a hurry to sign new long-term gas contracts. While it is possible that by the time the line is built traditional long-term contracts will cover all available capacity, the likelihood is that some part of the capacity – and possibly a large part – will remain uncontracted, at least on a long-term basis.

The link between these elements is as follows. The gas bubble comprises relatively low-cost gas. Within Russia, investment in both production and transmission infrastructure has largely been amortized. The only additional investment which is required within Russia is continued refurbishment of the transmission system, plus a certain amount of further investment in new capacity in the west of the country. If the projections of gas demand reduction in the Central region (made in Table 2.7) are correct, the cost of bringing these volumes to the Russian border with Belarus will be quite small.[1] If the demand reduction foreseen in the Central region proves overly optimistic, or if more of the gas from the bubble is needed for export, the next option would be to expand transmission capacity between the Urals – where demand reduction will be substantial – and the Russian border. The investment required for refurbishment and capacity expansion from the Urals would be

---

[1] Probably around $1 bn for one 56 inch line from Torzhok to Belarus; see Appendix to Chapter 1, Table A.4.

substantial – perhaps of the order of $4 bn.[2] Yet even taken together these investments are small – probably around 10–15% of the investment (estimated in Appendix to Chapter 1, Table A.5) required to bring comparable quantities of gas from Yamal Peninsula fields through new transmission lines to the Russian border.[3] Moreover, in terms of the additional revenues which could be earned from the increase in exports made possible by the additional transmission capacity, the payback time is probably not very great.

An important issue would be whether (and how) Gazprom will be able to finance such developments, either out of its own funds or through foreign borrowing. The participation of foreign companies, both as investors in facilities and/or as purchasers of gas, could be a crucial factor in providing confidence to the international lending community.

4.2 The consequences of the bubble for European gas markets

Assuming that the first Belarus–Poland line is built on schedule and brought up to full compression without delay, more than 30 BCM of Russian gas will be offered to European gas markets before the year 2000. As noted above, it is possible that all of these volumes will have been covered by traditional long term take-or-pay contracts by the time the pipeline is built, but the greater likelihood is that some part of them, possibly a large part, will not. Having made a significant investment in building the pipeline, there will be a strong incentive for Gazprom to use the capacity to the maximum possible extent.

---

[2] Assuming a 56 inch line with compression from (for example) Perm to the Belarus border.
[3] This is a very rough calculation because we are not comparing 'like with like'. In the case of the Yamal Project (see Appendix to Chapter 1, Table A.4) we have suggested that the total cost of delivering around 60 BCM of gas per year for several decades from the fields to the Belarus border is $36–8 bn. In the alternative case, the cost of creating 60 BCM of additional transmission capacity through which 'bubble' gas could be brought to the Belarus border is around $5 bn. However, we see only around 30–40 BCM of bubble gas available for a limited period of years. In addition, the Yamal Project set out in Table A.4 includes one pipeline (from the fields to Torzhok) for domestic use. Despite these caveats, the general argument holds good that a very large amount of capacity can be created at a fraction of the cost of new grassroots projects.

Therefore, it is probable that Gazexport, either on its own account or using the newly created joint venture marketing companies, will be seeking to sell parcels of gas on flexible contractual terms. The existence of the new marketing organizations which have been set up in most European countries will both give the Russians much better intelligence regarding market opportunities and margins in different parts of the gas chain than they have had in the past, and greatly weaken the ability of traditional buyers to oppose Russian proposals to increase gas sales in their markets.

The principal issues which are thrown up by the possibility of sales of Russian gas in European markets outside long-term contracts are access to pipelines and gas-to-gas competition. If Russian gas is to find its way to customers, it will need to do so through pipelines owned by the traditional 'club' of European gas companies. Since Gazprom now has collaborative trading houses or joint ventures with most of these companies, there may be no problem in persuading the latter to agree to move more gas through their networks, thereby increasing the market share for Russian gas.

However, this in turn raises two potentially problematic alternatives, both of which could set the stage for serious gas-to-gas competition taking place in Europe:

- Russian access to existing and new transmission facilities through which additional volumes may be moved, raising emotive (and commercially important) issues of the merchant status of European gas companies versus the possibility of their acting as 'transporters' of gas;
- sharing of merchant margins on new Russian gas supplies between Gazprom and European gas companies.

The outcome of negotiations on these questions is more likely to be a compromise than a confrontation, but a compromise under which some of the traditional rules of selling gas in European markets are broken. Confrontation is only likely if the Russians come to believe that their European partners are placing obstacles in the path of increased sales of Russian gas by failing to pursue available market opportunities, or by refusing to agree equitable access or revenue-sharing arrangements. Such a confrontation could prompt Gazprom to make alternative arrangements, in terms of both finding different partners in certain countries and building alternative infrastructure.

The principal uncertainty in these calculations is the degree of urgency on the part of Russian sellers to move potentially large volumes into European markets within a relatively short space of time. The expertise and experience of Gazprom/Gazexport in dealing with European gas markets suggest that 'dumping' of gas – defined as selling at prices which *substantially* undercut European border prices – is extremely unlikely. However, a decision fully to utilize the transmission capacity which has been created would probably result in pricing gas competitively with alternative gas supplies – as opposed to the traditional pricing against alternative fuels. Pricing gas against alternative gas supplies could lead to the acceleration of gas-to-gas competition with uncertain consequences for European gas prices, for traditional long-term contractual structures and for the possibility of developing new grass roots gas projects competitively. The impact on potential future supplies from non-Russian sources could be substantial. If a Russian gas bubble is perceived to be rolling into Europe in the second half of the 1990s and shows no sign of bursting for several years, it will require brave investors to go ahead with projects where the delivered cost of the gas is projected to be significantly in excess of $2 per mmbtu.

It is very important to stress that these circumstances have not been deliberately created by the Russians and will not necessarily be welcomed by them. It is possible that Gazexport will refuse to sell additional volumes of gas unless traditional long-term contracts can be signed. However, such decisions do not rest solely with the gas industry. Substantial political change is certain to occur in Russia over the next several years, but any government will have a significant requirement for foreign exchange earnings and therefore may place pressure on Gazprom (Gazexport) to increase exports of gas.[4] Under this kind of pressure, a careful strategy of gradually phasing in sales to European markets might become impossible. This is exactly the

---

[4] This is not to argue that such measures would necessarily produce greater revenues for Russia, only that Russian politicians may believe this to be the case.

type of commercial pressure which could greatly accelerate the existing forces in European gas markets – particularly in Germany – towards liberalization.[5]

4.3 Reacting to the bubble: security of supply issues revisited

Having advanced a Russian gas export scenario with potentially dramatic consequences for European gas markets, it is prudent to look at possible events which could significantly undermine all that has been said above. The principal events which could cause a serious disruption in gas trade are of a political and strategic nature. During the Soviet period, the character and likelihood of such events was extensively debated.[6] In the post-Soviet period, these events can be grouped under three main headings:

1   A catastrophic political or military event (or events) in Russia or Ukraine, or a deterioration in relations between the two countries, sufficiently serious fundamentally to disrupt the basis of the present gas trading relationship with Europe.
2   A deterioration in political relations between Russia and (East and West) European countries which brings security of gas supply back on to the agenda, not necessarily in the same context as in the Soviet era, but as a sufficiently serious issue to require government supervision and limitation of imports from Russia.
3   A resurgence of protectionism in European countries arising from the perception, or reality, of Russian gas taking market share from indigenous fuels (in particular domestically produced coal), or retarding the development of higher-cost European gas supplies (notably from Norway).

A military conflict within Russia or between Russia and Ukraine – even if this did not result in an immediate disruption of gas supplies to Europe –

---

[5] I have argued elsewhere that liberalization is much more likely to be driven by market pressures than by directives from the European Commission in Brussels. Moreover, significant liberalization of the German gas market would create enormous pressures for similar developments in other European countries. See Jonathan P. Stern, *Third Party Access in European Gas Markets: Regulation-Driven or Market-Led?* (London: Royal Institute of International Affairs, 1992).
[6] Bruce W. Jentleson, *Pipeline Politics: The Complex Political Economy of East–West Energy Trade* (New York: Cornell University Press, 1986), ch. 6.

would be sufficiently serious to prevent potential European customers from extending their exposure to this source of gas. Similarly, the arrival in Moscow of a blatantly anti-Western and anti-capitalist government might threaten even the continuation, let alone any expansion, of natural gas trade. Russia's European customers would seek to cut back their commitments and strengthen their existing security arrangements in anticipation of potential future disruptions. Such events would probably give rise to a complete reassessment of the merits of European market expansion based on Russian gas, and as such would have seriously adverse consequences for anticipated increases in European gas demand over the next two decades.

Consideration of a deterioration in the political relationship between Russia and Europe raises both questions concerning specific bilateral relationships between countries and more general questions concerning Russian–European relations. The former focus on Russia's relations with key transit countries, especially the Czech and Slovak Republics and also (with the building of the new pipeline corridor) Poland. Any major difficulties in these relationships would give rise to the same nervousness in Europe as has resulted from the difficulties between Russia and Ukraine.

On a more general level, a deteriorating political relationship between Russia and Europe – what President Yeltsin has termed a 'Cold Peace' – and protectionist sentiments have a number of elements in common. If political relations deteriorate, the general desire on the part of OECD Europe to expand trade with Russia may also fade, and this is likely to feed the latent European tendency towards protection of indigenous gas and energy production. In Central and Eastern Europe, a deterioration of relations with Russia would accelerate what are already strong inclinations to diversify gas supplies, despite the high costs involved. These events would not have the effect of decisively halting the flow of gas from Russia in the same way as the political/ military catastrophe scenario discussed above. But a deterioration in political relations with Russia would greatly complicate and could eliminate any prospect of significant expansion of gas trade.

From a European Union perspective, the Commission's position on security of natural gas supply is not particularly helpful in terms of future decision-making. The 1995 Green Paper on Energy Policy states:

Looking at the expected supply and demand picture for natural gas in the coming 20 years, it becomes clear that security of supply at competitive conditions should be a key goal of EC energy policy...

The Commission's position is that the growing interconnected European gas grid as well as the diversified gas infrastructure and sources of supply among Member States require that advantage is taken of the Community dimension to enhance security of supply. Short term security of supply in the gas sector requires a careful and in-depth examination of the specific measures necessary to respond to a gas supply crisis. The Commission believes that security of supply shall be ensured through an open market functioning under competitive conditions at all stages from production to transportation in conformity with the Treaty.[7]

This fails to address the likely trade-off between seeking the most competitively priced sources of gas and maintaining adequate diversity of supply, which is likely to face many countries if the scenarios of Russian gas supply foreseen in this study prove to be correct. If the majority of Europe's incremental gas requirement over the next 15 years is met by very competitively priced Russian gas supplies, this would significantly increase European dependence on Russia and therefore its vulnerability to any supply disruption.

A significant expansion of Russian exports which, under conditions of gas-to-gas competition, might occur at the expense of European supplies, as well as inhibiting future development of indigenous gas supplies, could meet with strong resistance from European producers. In such a situation, one might expect a revival of the early 1980s pressure for a limitation on the share of Russian exports destined for individual European countries.[8]

If European countries did try to place limits on Russian gas imports, it remains to be seen whether Russia would be able to use the transit provisions of the Energy Charter Treaty (see Box 4.1) to prevent such measures. The likelihood is that, if such measures were deemed to be in the national interest, European governments would be able to place informal political pressure on companies to prevent them from increasing Russian imports.

---

[7] *For a European Union Energy Policy*, Commission of the European Communities, Green Paper, COM(94) 659 final, 11 January, 1995, section 2.2, paras 46 and 47.
[8] For the background and details of the American sanctions and attempt to limit Soviet gas to 30% of demand in significant European markets see Jentlesen, *Pipeline Politics*.

Box 4.1 The Energy Charter Treaty: transit provisions

The Energy Charter Treaty, signed in Lisbon in December 1994, is a wide-ranging multilateral agreement which aims at fostering energy cooperation, and improving the conditions for energy trade, between the contracting parties, which include most OECD (including EU) member states, Central and East European countries, and former Soviet republics.[a]

Article 7 of the treaty deals with transit defined as:

the carriage through the area of a Contracting Party...of energy materials and products originating in the area of another state and destined for the area of a third state, so long as either the other state or the third state is a Contracting Party...[b]

The main principles are that:

Each Contracting Party shall take the necessary measures to facilitate the transit of energy materials and products consistent with the principle of freedom of transit and without distinction as to origin, destination or ownership of such...materials...or discrimination as to pricing on the basis of such distinctions, and without imposing any unreasonable delays, restrictions or charges.

Each Contracting Party undertakes that its provisions relating to transport of energy...shall treat...products in transit in no less favourable a manner than its provisions to treat such...products originating in or destined for its own area...

In the event that transit...cannot be achieved on commercial terms...the Contracting Parties shall not place obstacles in the way of new capacity being established...

A Contracting Party...shall not be obliged to:

– permit the construction...of facilities or
– permit new or additional transit through existing facilities

which it demonstrates...would endanger the security or efficiency of its energy systems, including security of supply.

A Contracting Party through whose area...products transit shall not in the event of a dispute over any matter...interrupt or reduce, permit any entity subject to its control to interrupt or reduce...the existing flow of energy...prior to the conclusion of the dispute resolution procedures.

A Contracting Party to a dispute may refer it to the Secretary General [who may]...appoint a Conciliator [who] shall seek the agreement of the parties to the resolution of a dispute...If within 90 days of his appointment he has failed to secure such an agreement, he shall recommend a resolution to the dispute...and shall decide interim tariffs and other terms and conditions to be observed for transit...

The extent to which these useful transit provisions will be enforceable remains to be seen. It is a pity that no precedent has yet been set in the use of the European Union Transit Directive, which might give some guidance on the settlement of such disputes.[c]

[a] This definition has been adapted from Robert De Bauw and Julia Doré, *The Energy Charter Treaty: Objectives, Expectations, Achievements and Prospects* (London: Royal Institute of International Affairs, 1995).
[b] Unless the two states decide otherwise and record their agreement with the Charter Secretariat.
[c] The Transit Directive requires EU member states 'to facilitate the transit of natural gas between high pressure transmission grids': *Official Journal of the European Communities*, no. L 147/37, 26 June 1990.

4.4 The bubble and after: scenarios to 2010

Looking ahead to 2010, the two most important questions for Russian gas exports are likely to be, first, the extent to which – and the period of time during which – the bubble of low-cost gas will be available to fill the additional transportation capacity which may be created during this period; and second, whether sufficient markets can be found in Europe for these volumes.

Russia intends to create an additional 30–40 BCM of export capacity by the year 2000, rising to 100 BCM by 2010. Given the size of the bubble at the end of 1994, the short-term constraint will principally be transportation rather than availability. In other words, until 2000 the availability of low-cost gas which, in normal times, would have been sold on the Russian domestic market at prices significantly lower than those being paid by European buyers will exceed Russia's ability to create transportation capacity to deliver these volumes to European markets. After the year 2000, the picture is likely to change, assuming the building of the second Belarus–Poland line and further expansion of the Ukraine transmission system, and the constraint may become availability rather than capacity.

European gas markets

The issue of the evolution of European gas demand is worthy of a separate analysis. However, for our purposes it is sufficient to note a variety of different views of demand in 2010 and volumes of gas yet to be contracted.

The principal difference between the projections in Table 4.1 is in their projections of gas demand in power generation for individual countries, and for Germany in particular. The principal caveat to be entered about these views of the European market is that they have all been constructed using current gas pricing methods and contractual structures. This author is not aware of any scenario which takes into account the possibility of gas-to-gas competition and capturing market share by pricing gas significantly below competing fuels (specifically gas oil). Some demand forecasts also show the amount of demand available for 'new projects', although this inevitably includes assumptions about renewals and extensions of existing contracts. Typical European gas company 'yet to be contracted' forecasts for 2010 have figures in the range of 86–130 BCM.

Table 4.1 Projections of European gas demand to 2010 (BCM)[a]

	1993	2000	2010
OECD[1]	352.1	406–428	470–502
Central/Eastern[1]	72.2	86–96	118–134
Total Europe[2]			570–600
OECD[3]		364	470
Western Europe[4]			428
Eastern Europe[4]			107
Western Europe[5]		383–430	431–516
Western Europe[6]			432–458
Western Europe[7]			535
Central/Eastern Europe[7]			101
Western Europe[8]			428–492
Central/Eastern Europe[8]			107–128

[a] All the figures in this table, excepting those derived from source [2], have been increased by 6.9% from the original source to make them compatible with the Russian cubic metres used in the rest of the study.

[1] *The Development of International Gas Markets* (Paris: Cedigaz, 1994).

[2] Gazprom forecast from R. I. Vyakhirev, 'The Russian Gas Industry in the Context of the Russian and World Economy', paper presented to the Conference on 'Natural Gas: Trade and Investment Opportunities in Russia and the CIS', Royal Institute of International Affairs, London, 13–14 October, 1994.

[3] International Energy Agency, *World Energy Outlook* (Paris OECD, 1994).

[4] Statoil estimates; Karen Fossli, 'Western Europe Gas Needs to Rise 60%', *Financial Times*, 4 November, 1993.

[5] Shell International Gas.

[6] Ruhrgas forecast from Gerhard Enseling, 'Analysing Future Sources of Gas Supply for Europe', paper presented at the European Gas Strategies '95 Conference, Amsterdam, February 1995.

[7] Purvin and Geertz forecast from 'European Gas Demand Will Be Strong to 2010; Costs Cloud Supply Picture', *Oil and Gas Journal*, 16 May, 1994, pp. 32–52.

[8] SNAM forecast from Dominico Dispenza, 'Europe's Need for Gas Imports Destined to Grow', *Oil and Gas Journal*, 13 March, 1995, pp. 45–8.

The Gazprom projection of the European gas 'deficit' in 2010 is of the order of 200 BCM.[9]

---

[9] R.I. Vyakhirev, 'The Russian Gas Industry in the Context of the Russian and World Economy', paper presented to the conference on 'Natural Gas: Trade and Investment Opportunities in Russia and the CIS', Royal Institute of International Affairs, London, 13–14 October, 1994, slide 2.

Focusing on the aspect of these projections which is important for our purposes, if the 'yet to be contracted' figure for 2010 is around 100 BCM per year, then the Russian availability which we have foreseen would be sufficient to satisfy all incremental demand in Europe. For a variety of reasons – not least the security issues mentioned above – it seems unlikely that Russian gas would be allowed to monopolize incremental demand in this way over such a long period of time. However, if 'yet to be contracted' gas demand in Europe is closer to the 200 BCM per year suggested by Gazprom, then there should be less of a problem in accommodating an additional 100 BCM of Russian gas by 2010. It is important not to lose sight of other likely sources of supply for the European market around the turn of the century – pipelines carrying additional British and Norwegian supplies promise to be completed around the same time as the Belarus–Poland line; on the other hand, some are foreseeing problems with Algerian supplies, given the political instability in that country.

Scenarios to 2010

Looking beyond the gas bubble, in Table 4.2 four scenarios have been constructed for the year 2010. These scenarios look at the Russian gas balance from the somewhat unusual standpoint of giving the markets for gas – export and domestic – the position of principal importance in determining the rest of the balance. Two different variants are advanced for both export volumes and domestic demand.

Export markets are divided between the former Soviet republics and Europe. Exports to former Soviet republics are assumed to be held flat at 80 BCM for the entire period. While these countries are not assumed to be paying the same prices as European customers, even by 2010, their importance as transit routes will guarantee them a continuing level of deliveries similar to their present imports.[10] For European markets, the variants are based on different scenarios of how quickly those markets might be able to accommodate large increments of Russian gas supply.

The development of Russian internal demand is the most difficult element in the balance. Two variants are suggested here. Under the high-demand

---

[10] The inter-governmental agreements noted in Chapter 3 above allow for significant increases in deliveries to Ukraine and Belarus. The assumption here is that a combination of lack of demand and inability to pay will result in no significant increase in Russian exports.

Table 4.2 Russian gas balance: scenarios to 2010 (BCM)

	1994[a]	2000	2010			
			A	B	C	D
Exports						
to FSU	79	80	80	80	80	80
to Europe	106	140	150	200	150	200
Demand	359	320	350	350	400	400
Production[b]						
actual	607					
capacity	640	640	640	640	640	640
Pipeline fuel	58	60	60	60	60	60
'Bubble'	33	40				
or shortfall[c]			0	55	55	110
Options for additional supply						
1. Domestic production						
Yamal fields			1–2 lines		3–4 lines	
Shtokmanovskoye				30		30
2. Central Asian 'imports'						
Turkmenistan		20		20		20
Kazakhstan		10		30		30

[a] Estimate including 4 BCM imports and 9 BCM net additions to storage.
[b] At existing fields and satellites.
[c] Gas requirement plus 10% pipeline fuel.

outcome, demand begins to increase rapidly again, reaching its 1990 level of 400 BCM around 2010. This is a scenario of moderate economic growth with some economic restructuring towards energy-efficient industries. However, tolerance of inefficient industries remains and internal gas prices – particularly for households – stay significantly below market levels. Under the low-demand outcome, a highly dynamic Russian economy is envisaged with fast economic restructuring, substantial replacement of inefficient industrial capacity and elimination of subsidized pricing. Despite rapid economic growth, plant replacement and efficiency measures prevent industrial gas and power demand from increasing, and residential/commercial demand is held down by increasing prices to market levels. Demand increases somewhat from the 2000 level of 320 BCM to reach 350 BCM in 2010.

Of course, these are not the only possible outcomes. However, this author believes they span the likely boundaries of demand in 2010. In particular, it is very difficult to see demand exceeding 400 BCM at that date unless something akin to central planning returns in Russia.

The aim in advancing the scenarios shown in Table 4.2 is to look at the ease with which increases in exports can be accommodated in terms of additional production capacity which may be required in the period up to 2010. For purposes of simplicity, use of gas by the pipeline system has been assumed at 9.5% of production throughout the period; changes in storage and losses have been ignored. The principal production assumption is that the capacity of existing fields plus satellites remains at 640 BCM per year throughout the period. The table shows that the bubble of available gas increases slightly from 33 BCM in 1994 to 40 BCM in the year 2000, *despite the fact that exports will have risen by 35 BCM during this period*. To the extent that exports fail to fill the transmission capacity which has been created by 2000, the bubble will be larger than 40 BCM.

For the year 2010, four scenarios are advanced:

- low exports/low demand;
- high exports/low demand;
- low exports/high demand;
- high exports/high demand.

The consequences in terms of additional production capacity are that if exports grow by only 50% over the next 15 years and demand remains low (around the 1994 level), no additional production capacity is required in 2010. If exports double by 2010 or demand returns to its 1990 level, then 55 BCM of additional production will be required. Only if exports double *and* demand returns to the 1990 level will an additional 110 BCM of production be required by 2010.

Supply options
Table 4.2 also suggests two different sets of options for meeting additional supply requirements, based respectively on domestic production and on 'imports' from Central Asia.

The conventional option is to bring gas from the Yamal fields, or the Shtokmanovskoye field in the Barents Sea, on-stream after 2000. The volume of 55 BCM means that only two lines from the Yamal Peninsula would be required, or one line plus gas from the Barents Sea. A 110 BCM requirement would need three or four lines from Yamal, or two or three lines plus Shtokmanovskoye.[11]

There are some alternatives to the domestic gas supply options. As noted in Chapter 3, infrastructure exists to bring Central Asian gas supplies from Turkmenistan and Kazakhstan into Russia at relatively low cost. The infrastructure which brought around 40 BCM of Turkmen gas to Russia and Ukraine in 1990 is currently underutilized, and will probably remain underutilized until 2000, by around 20 BCM per year. The Karachaganak gas field lies less than 300 km from the existing Orenburg pipeline which – with the depletion of the Orenburg field – could eventually allow up to 30 BCM of Kazakh gas to be brought to Europe after 2000.

Thus around 50 BCM of Central Asian gas can be brought into Europe with very little more than refurbishment expenditures on existing infrastructure. Central Asian gas can either be delivered to Europe by displacement – which will undoubtedly be the preferred option of sellers in both Turkmenistan and Kazakhstan – or sold in Russia and Ukraine, thereby freeing up more Russian gas for sale in Europe (almost certainly the preferred Russian option). The eventual outcome would need to be a compromise between these positions. Russia would also need to take into account that Central Asian countries might try to use the transit provisions of the Energy Charter (see Box 4.1) to market their gas in Europe.

Two footnotes should be added to the Central Asian options: first, that a somewhat smaller volume of gas would be available for the year 2000, adding another 30 BCM to the bubble of 40 BCM which already exists in that year; second, that given the established reserve base in both Turkmenistan and Kazakhstan, plus established transportation routes from both countries through Russia, significant expansion of these supply options would be

---

[11] In fact it seems unlikely that the Yamal Peninsula fields will be developed unless the supply requirement is at least 80 BCM/year because of the need to build (at least) a three-line development from the fields to Ukhta (see Map 1.3 and Table A.4, Appendix to Chapter 1).

possible at a relatively low cost, significantly lower than the cost of either transporting gas from the Yamal Peninsula, or transporting Central Asian gas to Europe via exotic non-Russian routes.

If the Central Asian options are chosen, then for scenarios A-C only 5 BCM of additional Russian production would be required in 2010; for scenario D (high exports/high demand) 60 BCM of additional production would be required.

4.5 Possible Gazprom strategies to 2010

Analysis of Gazprom's likely gas markets – domestic and export – suggests that the current emphasis on creating major new supply projects from the Yamal Peninsula (and elsewhere) over the next decade is misplaced. Unless both demand and exports reach high levels very quickly, existing fields plus satellites can meet all demand up to 2005. For 2010 the least-cost way of meeting most demand scenarios will be for Russia to import gas from Central Asia through pipelines which already exist and require only refurbishment expenditures. Only in a situation where both internal demand *and* exports reach high levels will significant additional production be required before 2010.

In its strategies for meeting anticipated gas requirements, Gazprom should be evaluating a wide range of options of which the Yamal fields seem to be the most expensive on the supply side; and the demand-side measures seem to have relatively low economic, but possibly high political, costs. To the extent that internal Gazprom politics and a desire to maintain employment levels point in the direction of continuing to pursue the opening up of the Yamal fields, this could further inflate and extend the life of the anticipated gas bubble.

Much more important for Gazprom will be to manage a major and rapid culture change whereby the organization moves its focus from opening up new fields and building new transmission lines to devising new strategies for its three major markets: the Russian market, the market in the former Soviet republics and the export market in Europe.

The Russian market

Here the biggest issues for Gazprom to deal with are non-payment and prices. Neither of these issues is within Gazprom's control, but it can try to continue to apply pressure on the central and regional political authorities to permit disconnection of (or at least reduction in deliveries to) industrial enterprises which are not paying bills. At the same time, it can press for real price increases on the basis that – even if new government taxation removes the revenue benefits from higher prices – the creation of greater conservation and efficiency incentives will exert downward pressure on demand and thereby reduce future supply costs.[12] The magnitude of the culture change required for Gazprom to view reducing the need for additional supply as a desirable strategy should not be underestimated.

The FSU republic markets

Here also the two biggest issues are non-payment and prices. The crucial need in the former republics is to stabilize interdependence relationships where deliveries of gas are traded off against unimpeded transit of gas to European countries. Gazprom's strategy here has been to regain control of facilities by offering gas at reduced prices in return for equity ownership of transportation assets. This appears to have been successful in Belarus and Moldova but not yet in Ukraine, where the issue of non-payment is most acute and the transit vulnerability is greatest. The only viable future strategy will be to try and move deliveries to a credible and workable commercial basis, avoiding confrontations which will not benefit either side and may do long term commercial and political damage to European trading relationships.

European export markets

Here Gazprom's freedom of action is greatest, but its choices are less straightforward. There are two strands to its current strategy: moving downstream to share the profits, particularly in transmission and storage, and creating a second export corridor through Belarus and Poland. The problem for Gazprom will be to decide whether to maximize revenues or maximize (and protect) market share.

---

[12] However, price increases will be of limited use – and may even be counterproductive – for as long as non-payment continues to any significant extent, particularly in the industrial sector. Non-payment in the residential market may continue to be tolerated until some form of metering can be installed. But this will give rise to major problems as household demand increases.

In Central and Eastern Europe, the attitude seems to be one of protecting what has until now been a captive set of markets, which other suppliers have been very slow to approach. In the future the issue may be the extent to which Gazexport, in the renegotiation of contracts, wishes to give these countries serious commercial incentives to reinforce the political and strategic desire to look elsewhere for gas supplies.

In OECD Europe the major issue is likely to be how quickly, and on what contractual basis, additional gas can be sold through the new capacity currently being built. The response of Gazexport will be crucial if – as seems likely – it has failed to sell some part of these volumes under traditional long-term contracts by the time the pipeline is built. One aspect of this situation which will quickly become prominent will be whether the Belarus and Polish sections of the pipeline can be financed in the absence of traditional long-term contracts. If the financial community is unwilling to go ahead in the absence of such contracts, then Gazprom and other potential investors will need to come forward with the money themselves.[13] Assuming the line is financed and proceeds more or less on schedule, there will be a strong imperative to pass as much gas as possible through it at the earliest possible opportunity in order to recoup the considerable financial outlay.

This imperative will speak in favour of rapid sales, and an important part of Gazprom's strategy will be how such sales should be handled. The first step will be to see how much gas will be contracted in Germany by the Gazprom/Wintershall trading house Wingas, and by Gazexport's other traditional customer – Ruhrgas. If gas remains available when the possibilities of the German market have been exhausted (which seems likely), then other trading houses in Eastern and Western Europe may be approached to seek markets for the balance of volumes. In addition, when indications of availability and price become clearer, the possibilities should not be excluded that parties in Germany or other countries might approach Gazexport directly. Such parties would probably be independent power generators (IPPs) with an interest in a direct purchase of Russian gas. IPP aspirants will be interested in opportunities in a range of countries, including Britain, where, it should not

---

[13] The postponement of the sale of Gazprom's equity to foreign companies, announced in March 1995, has temporarily closed the option of raising foreign exchange by this means.

be forgotten, Gazprom has recently taken a 10% stake in the 'Interconnector' pipeline to be built between Britain and Belgium.[14]

Gazprom's key decision, therefore, is on the best way to expand European export markets into which to feed its gas bubble. If it stays within the present rules of the European gas business – merchant sales contracts, pricing gas against alternative fuels, no gas-to-gas competition – then the volumes which it sells may be rather limited in comparison to the gas and capacity which will be available. If, however, it begins to break those rules, the outcomes become extremely uncertain in terms of the possible response of other European suppliers and hence the revenues that Gazprom can expect. Despite that uncertainty, if markets for Russian gas fail to open up in Europe over the next five years, the temptation to compete strongly on price to create them may be very great.

---

[14] This would give Gazprom around 1.5 BCM of capacity in the line: 'Interconnector Comes Alive but Free Trade Stays in Coma', *Gas Matters*, December 1994, pp. 1–8.